Praise for RESILIENCE

"When the next disruption strikes, some will fall—and some, following the lessons of this book, will rise."

> —Juan Enriquez, author of *As the Future Catches You* and *Homo Evolutis* and managing director of Excel Venture Management

"Spending time with Andrew Zolli's mind—that is what you will experience when reading *Resilience*—provides an understanding of the deep structures that will govern success in the coming century."

> —Bruce Mau, cofounder and director of Massive Change Network

"Resilience is the most important key to healing a planet that faces the most dangerous of times. More important and far more essential than either sustainability or corporate responsibility. Andrew Zolli and Ann Marie Healy's new book has arrived at a time when we need their insight and wisdom most. Understanding resilience is imperative for our very health and survival."

> —Jeffrey Hollender, cofounder of Seventh Generation and founder of Jeffrey Hollender Partners

RESILIENCE

Why Things Bounce Back

ANDREW ZOLLI
& ANN MARIE HEALY

FREE PRESS

New York   London   Toronto   Sydney   New Delhi

*f*P

Free Press
A Division of Simon & Schuster, Inc.
1230 Avenue of the Americas
New York, NY 10020

First Free Press hardcover edition July 2012

FREE PRESS and colophon are trademarks of Simon & Schuster, Inc.
For information about special discounts for bulk purchases, please contact
Simon & Schuster Special Sales at 1-866-506-1949
or business@simonandschuster.com.

The Simon & Schuster Speakers Bureau can bring authors to your live event.
For more information or to book an event contact the
Simon & Schuster Speakers Bureau at 1-866-248-3049
or visit our website at www.simonspeakers.com.

Designed by Ruth Lee-Mui

Manufactured in the United States of America

3   5   7   9   10   8   6   4   2

Library of Congress Cataloging-in-Publication Data

Zolli, Andrew.
Resilience : why things bounce back / Andrew Zolli & Ann Marie Healy.
p. cm.
Includes bibliographical references and index.
1. Fault Tolerance (Engineering) 2. Stability 3. Hysteresis
4. Resilience (Personality Trait) I. Healy, Ann Marie. II. Title.
TA169.Z65 2012
303.4—dc23   2011051818

ISBN 978-1-4516-8380-6
ISBN 978-1-4516-8384-4 (ebook)

For Emilia, Benjamin, Nolan, and Evelyn
May yours be a more resilient future

CONTENTS

RESILIENCE

INTRODUCTION

THE
RESILIENCE IMPERATIVE

On January 31, 2007, the long, narrow alleys and wide boulevards of Mexico City were filled with typical early morning sounds: children running through open doors; families preparing for the day; and street vendors cooking up tortillas, one of Mexico's main food staples.

Yet this was to be no ordinary day. On this day, the price of corn—the main ingredient in tortillas—would hit an all-time high of 35 cents a pound, a price that would have been unfathomable just a year before. Corn was suddenly 400 percent more expensive than it had been just three months earlier. With half of all Mexicans living below the poverty line, a sudden increase of this magnitude was not just a nuisance, it was a potential humanitarian and political crisis.

As the sun lifted higher in the sky, the voices of tens of thousands of citizens, farmers, and union activists could be heard gathering in one of the city's central squares. Above their heads, they raised not weapons, but ears of corn. The tortilla riots, as they came to be called, echoed throughout the day, taking over one of the main downtown streets and challenging the new government of President Felipe Calderón. Well into the evening, protestors chanted, *"Tortillas sí, pan no!"*—a pun on Calderón's National Action Party, the PAN, which also means "bread" in Spanish—and barked out their suspicions about just who was behind the rise in prices: the government, big business, and the wealthy elite of the country. Union leaders and television celebrities railed against corporations for price fixing and chastised the beef and pig ranchers for hoarding their grains.

While the ranchers and political leaders were natural objects of class indignation, they were not, this time at least, the principal culprits. Indeed, the protestors could scarcely have guessed the truth: The slowly burning fuse that had ignited the explosion in corn prices had been lit several years before and a thousand miles away by a seemingly disconnected event—Hurricane Katrina.

Here's how: In August 2005, the impending winds of the devastating hurricane had prompted the mass evacuation and shutdown of the 2,900 oil rigs that dot the Gulf Coast from Texas to Louisiana, disrupting almost 95 percent of oil production in the Gulf for several months. In the aftermath of the storm, the price of gasoline in America surged, in some places by as much as 40 cents per gallon in a single day. This spike in oil prices made corn—the primary ingredient in the alternative fuel ethanol—look relatively cheap by comparison and spurred investment in domestic ethanol production. U.S. farmers, among the most efficient and most heavily subsidized in the world, were encouraged to replace their edible corn crops with inedible varieties suitable for ethanol production. By 2007, even Congress had gotten in on the act, mandating a fivefold increase in biofuel production—with more than 40 percent of it to come from corn.

Amid the euphoria of this ethanol investment bubble, almost no one considered potential impacts on Mexico's peasant farmers, who, in the decade between the passage of NAFTA and the arrival of

Katrina, had found themselves thrust into international competition with powerhouse U.S. agribusinesses north of their border. American corn growers routinely sold (many would argue dumped) their product on Mexican markets at almost 20 percent less than it cost to produce it. Unable to keep up—even with the support of their own domestic subsidies—many rural Mexican farmers had switched the variety of corn they grew, switched crops altogether, or abandoned their farms, swelling the ranks of Mexico City's underclass and further accelerating Mexico's position as a primary market for cheap U.S. varieties.

As NAFTA took hold, this expanding corn import market had also become increasingly dominated by a tiny clique of powerful transnational corporations, mostly headquartered in the United States, including Cargill and Archer Daniels Midland, along with their Mexican subsidiaries. These companies accelerated the transition already at work by doing what dominant incumbents instinctively do: concentrating power, tightening their control over the market, and squeezing out smaller suppliers. The result: Mexico, famous as the place that domesticated the growing of corn ten thousand years ago, soon became a net food importer—and the third largest of U.S. agricultural products—much of it channeled through a tiny constellation of companies.

It was against this backdrop that, in the year that followed Katrina, with increasing amounts of the United States' domestic supply being diverted to ethanol, the price of corn became inextricably coupled to the price of oil—not only because ethanol and oil are comparable fuels, but also because it takes an enormous amount of petroleum-derived fertilizers to grow corn in the first place. As the price of a barrel of petroleum fluctuated, the price of a bushel of corn began increasingly to move in lockstep. When global speculation drove the cost of a barrel of oil to nearly $140, the now-linked price of corn also skyrocketed, provoking what may become an archetypal experience of the twenty-first century: a food riot.

•   •   •

We are, of course, used to these kinds of stories. Each week, it seems, brings some unforeseen disruption, blooming amid the thicket of overlapping social, political, economic, technological, and environmental systems that govern our lives. They arrive at a quickening yet erratic pace, usually from unexpected quarters, stubbornly resistant to prediction. The most severe become cultural touchstones, referred to in staccato shorthand: Katrina. Haiti. BP. Fukushima. The Crash. The Great Recession. The London mob. The Arab Spring. Other nameless disruptions swell their ranks, amplified by slowly creeping vulnerabilities: a midwestern town is undone by economic dislocation; a company is obliterated by globalization; a way of life is rendered impossible by an ecological shift; a debt crisis emerges from political intractability. If it feels like the pace of these disruptions is increasing, it's not just you: It took just six months for 2011 to become the costliest year on record for natural disasters, a fact that insurance companies tie unambiguously to climate change. Volatility of all sorts has become the new normal, and it's here to stay.

While the details are always different, certain features of these disruptions are remarkably consistent, whether we're discussing the recent global financial crisis, the geopolitical outcomes of the war in Iraq, or the surprising consequences of a natural disaster. One hallmark of such events is that they reveal the dependencies between spheres that are more often studied and discussed in isolation from one another. The story of the tortilla riots, for example, makes visible the linkages between the energy system (the oil rigs) the ecological system (Katrina), the agricultural system (the corn harvest), the global trade system (NAFTA), social factors (urbanization and poverty), and the political systems of both Mexico and the United States.

We tell such stories to encourage humility in the face of the incomprehensible complexity, interconnectivity, and volatility of the modern world—one in which upheavals can appear to be triggered by seemingly harmless events, arrive with little warning, and reveal hidden, almost absurd correlations in their wake. Like pulling on an errant string in a garment, which unravels the whole even as it reveals how the elements were previously woven together, we make sense of these stories only in retrospect. Even with a deep understanding of the individual systems involved, we usually find it difficult to untangle the chain of causation at work. And for all of the contributions of the much-ballyhooed Information Age, just having more data doesn't automatically help. After all, if we could actually see each of the individual packets of data pulsing through the Internet, or the complex chemical interactions affecting our climate, could we make sense of them? Could we predict in detail over the long term where those systems are headed or what strange consequences might be unleashed along the way? Even with perfect knowledge, one can't escape the nagging suspicion we're ballroom dancing in the middle of a minefield.

So what to do?

If we cannot control the volatile tides of change, we can learn to build better boats. We can design—and redesign—organizations, institutions, and systems to better absorb disruption, operate under a wider variety of conditions, and shift more fluidly from one circumstance to the next. To do that, we need to understand the emerging field of resilience.

Around the world, in disciplines as seemingly disconnected as economics, ecology, political science, cognitive science, and digital networking, scientists, policymakers, technologists, corporate leaders, and activists alike are asking the same basic questions: What causes one system to break and another to rebound? How much change can a system absorb and still retain its integrity and purpose? What characteristics make a system adaptive to change? In an age of constant disruption, how do we build in better shock absorbers for ourselves, our communities, companies, economics, societies, and the planet?

Like a developing Polaroid, the insights, lessons, and rules of thumb they are discovering are revealing an entirely new field—a set of generalizable insights for building social, economic, technical, and business systems that anticipate disruption, heal themselves when breached, and have the ability to reorganize themselves to maintain their core purpose, even under radically changed circumstances.

With this in mind, consider how the Mexicans might have been spared their difficulties. Larger stockpiled reserves of corn, more diversified food crops, better real-time data, and better modeling of the impacts of U.S. corn crop diversion might obviously have helped; so too might a mechanism to rapidly secure alternative suppliers in a crisis, or restructuring the market to dampen the monopolies' power, or investments in social programs for the poor to mitigate the effects of the price spike. Or one might just as readily have intervened in another point in the causal chain—say, by diversifying U.S. energy production—so that even a major hurricane wouldn't spur the diversion of corn to ethanol production in the first place.

The strategies implied in each of these interventions—ensuring that there are sufficient reserves available to any given system; or diversifying its inputs; or collecting better, real-time data about its operations and performance; or enabling greater autonomy for its constituent parts; or designing firebreaks so that a disturbance in one part does not disrupt the whole—are, at their core, strategies of resilience. As we'll see, they can be applied at any scale, from whole civilizations to communities and organizations, to the lives of individual people.

Defining resilience more precisely is complicated by the fact that different fields use the term to mean slightly different things. In engineering, resilience generally refers to the degree to which a structure like a bridge or a building can return to a baseline state after being disturbed. In emergency response, it suggests the speed with which critical systems can be restored after an earthquake or a flood. In ecology, it connotes an ecosystem's ability to keep from being irrevocably degraded. In psychology, it signifies the capacity of an individual to deal effectively with trauma. In business it's often used to mean putting in place backups (of data and resources) to ensure continuous operation in the face of natural or man-made disaster. Though different in emphasis, each of these definitions rests on one of two essential aspects of resilience: continuity and recovery in the face of change.

Throughout this book, we will explore resilience in both systems and people. Accordingly, we frame resilience in terms borrowed from both ecology and sociology as *the capacity of a system, enterprise, or a person to maintain its core purpose and integrity in the face of dramatically changed circumstances.*

To see what we mean, let's explore a metaphor used widely in resilience research. Imagine for a moment you are overlooking a vast landscape of imaginary hills and valleys, stretching out in every direction. Like something from a Borges fantasy, each valley in this panorama presents a significant variation on your present circumstance, an alternative reality with its own unique characteristics, opportunities, resources, and dangers. Each hill in this landscape can be thought of as the critical threshold or boundary separating these worlds—once you pass its peak, you will, for good or ill, inexorably roll into the adjacent existential valley below. In some of these new circumstances, you may find life quite easy; in others, you may find things challenging; and in a few you may find your new reality so difficult that adaptation is all but impossible.

As in real life, any number of sudden and serious disruptions might cause you to be "flipped" over the threshold separating your present context and a new one: Perhaps you experience a flood, or a drought, an invasion, or an earthquake, or perhaps your valley becomes too sparsely populated or too crowded to occupy. Perhaps your business encounters an economic or energy shock, a technological or competitive shift, a sudden shortage of raw materials, or the pricing in of environmental factors that were previously unaccounted for. Unfortunately, many of these thresholds may be crossed only in one direction: Once forces have compelled you into a new circumstance, it may be impossible for you to return to your prior environment. You'll have entered a new normal.

To improve your resilience is to enhance your ability to resist being pushed from your preferred valley, while expanding the range of alternatives that you can embrace if you need to. This is what resilience researchers call *preserving adaptive capacity*—the ability to adapt to changed circumstances while fulfilling one's core purpose—and it's an essential skill in an age of unforeseeable disruption and volatility.

There are, of course, many ways to expand your range of habitable niches. You could reduce your material needs in order to subsist in more resource-poor settings; you could learn to use a wider array of resources, so you could survive, MacGyver-like, on whatever might be locally available; you could invent a new technology, liberating yourself from a traditional constraint; you could modify tools designed for one niche to suit another; or you could learn to collaborate with the local denizens so that you don't have to go it alone.

As it is for people, so it is for systems, businesses, nations, and even the planet as a whole—all can occupy a number of different, stable states, some vastly preferable to others. Planetary boundaries (so named by the resilience researcher Johan Rockström and his colleagues at the Stockholm Resilience Centre) are thresholds that keep the entire biosphere from flipping, suddenly and potentially catastrophically, into a new state: They include factors like the acidification of the oceans, the loss of biodiversity, human transformation of the land, and the availability of clean water. Of the nine thresholds Rockström's team has identified, three are currently exceeded; another four are approaching their limit. Like Russian nesting dolls, these planetary boundaries set the limits and context for all human activities, from settlement and migration to conflict and commerce, and spur the development of new forms of technology and exchange.

Enhancing the resilience of an ecosystem, an economy, or a community can be achieved in two ways: by improving its ability to resist being pushed past these kinds of critical, sometimes permanently damaging thresholds, and by preserving and expanding the range of niches to which a system can healthily adapt if it is pushed past such thresholds.

In principle, there are as many ways for a complex system to adapt as there are circumstances for it to adapt *to*. However, the dynamics of our present era—including the relentless quest for organizational efficiencies, the deep stressing of ecological systems, and the interconnections that bind us all—move certain approaches to the forefront. These patterns, themes, and strategies appear again and again, in large ways and small, wherever resilience is found.

PATTERNS OF RESILIENCE

From economies to ecosystems, virtually all resilient systems employ tight *feedback mechanisms* to determine when an abrupt change or critical threshold is nearing. As we'll see in the next chapter, in an ecosystem like a coral reef, certain species' behavior can change to prevent a system from flipping into a degraded state. In a human context it's much the same, though when people do it, we're often supported by an array of tools and technologies that provide us with a greater sense of situational awareness.

For example, the venerable check engine light on your car's dashboard, if attended to, can be thought of as maintaining the resilience of the engine (and hopefully, the driver) by helping you understand that something is wrong under the hood and encouraging you to quickly get to a mechanic. In a much more sophisticated but analogous way, we're now in the midst of a massive real-time instrumentalization of many human systems, from health care to business operations and international development. We're soaking the world in sensors, and the feedback data that these sensors produce is a powerful tool for managing systems' performance and amplifying their resilience—particularly when those data are correlated with data from other such systems.

The U.S. Geological Society, for example, is building a tool called the Twitter Earthquake Detector (TED) that links its seismometers to the social media service. When a quake is detected, the TED instantly scans for tweets about the location and severity of damage and maps them geographically, enabling faster and more targeted disaster response. Similarly, in sub-Saharan Africa, health researchers are building powerful predictive models of disease outbreaks by studying migration patterns derived from people's cell-phone usage; by studying where they are calling, they deduce where they are moving, and can assign medical resources to where they will be needed in the future, not just where they are needed in the present. These same researchers found that they can determine the economic well-being of citizens by passively studying the denominations in which they buy airtime for their cell phones. Buying airtime in $1 increments is a sign of greater well-being than buying airtime in 10-cent increments; a sudden, downward shift in buying habits may be an early warning sign of an impending economic disruption. All of these approaches rest on sorting, sifting, and combining open, real-time data from vast sensor networks and using them to create meaningful feedback loops.

When such sensors suggest a critical threshold is nearing or breached, a truly resilient system is able to ensure continuity by *dynamically reorganizing* both the way in which it serves its purpose and the scale at which it operates. Many resilient systems achieve this with embedded countermechanisms, which lie dormant until a crisis occurs. When that happens, they are dispatched, like antibodies in the bloodstream, to restore the system to health.

Another way to bolster a system's resilience is to *de-intensify* or *decouple* the system from its underlying material requirements or to diversify the resources that can be used to accomplish a given task. Under duress, some resilient systems may even detach themselves entirely from their larger context, localizing their operations and reducing their normal dependencies.

For example, many global companies have awakened to the fact that we are approaching a critical threshold concerning reliable access to water, driven by the competing needs of agriculture, human consumption, and industry. Sustainability experts at Nike recently calculated that it takes a whopping 700 gallons of water to produce a single organic cotton T-shirt. (Think about that the next time you stand in front of a wall of three-dollar tees at Walmart.) It's little wonder that they, and others in their industry, are now aggressively pursuing efforts to develop less water-intensive approaches to production and manufacture—for example, by using less water to grow cotton and dye textiles. They're trying to decouple water from apparel, to the greatest degree possible.

This kind of reorganization is made feasible by certain structural features of resilient systems. While these systems may appear outwardly complex, they often have a simpler internal *modular structure* with components that plug into one another, much like Lego blocks, and—just as important—can unplug from one another when necessary. This modularity allows a system to be reconfigured on the fly when disruption strikes, prevents failures in one part of the system from cascading through the larger whole, and ensures that the system can scale up or scale down when the time is right.

Herb Simon, the polymathic psychologist, political scientist, economist, and computer scientist, demonstrated the importance of this kind of modularity with a famous parable about two watchmakers, Hora and Tempus. Both craftsmen built watches of equal complexity and beauty, comprising hundreds of parts. Yet Hora's business flourished, while Tempus's business failed.

The reason? Hora built his watches modularly, fitting individual components into hierarchical assemblies that could be snapped together to complete the whole. Tempus, on the other hand, simply built his watches piece by piece until each was completed.

In Simon's parable, the watchmakers are occasionally interrupted by a phone call with more orders for watches. When that happens, they must restart the task they were doing prior to the interruption. As such, Tempus must begin each watch over and over again, while much of Hora's prior work is preserved. And boy, does it make a difference: If both watchmakers are interrupted just 1 percent of the time, Hora will complete 9 watches for every 10 he tries to build. Tempus, on the other hand, will finish just 44 watches for every 1 million.

To encourage this beneficial modularity, many resilient systems are *diverse at their edges* but *simple at their core*. Think of the DNA in a cell, or the communications protocols governing the Internet: These specialized languages encode a vast menagerie of inputs and outputs, yet as protocols, they remain utterly basic, evolving slowly, if at all. The electrical grid, for example, in effect translates power generated from a number of sources—from nuclear power plants to windmills—into countless useful forms of work. At the center of this vast machine is an unchanging language of currents, voltages, and electrons. The resilience of the overall power system is improved as we expand the diversity of sources that feed it and improve the efficiency of the tasks that we use the resulting electricity for, yet the underlying core protocol of the system remains unchanging. (The converse is also true: Like the Mexican food system, the resilience of the power system is reduced as we narrow the diversity of sources that feed it.)

This modularity, simplicity, and interoperability enable the components of many resilient systems to *flock* or *swarm* like starlings when the time is right and to break into islands when under duress. These are the very features that make things like cloud computing possible—in which groups of linked, redundant servers swarm together, scaling up and down to complete a given task, then disband. Similarly coordinated approaches to resilience are found in realms as seemingly disparate as bacteria and the battlefield.

Yet this kind of modular distributed structure is only part of the story. Paradoxically, resilience is often also enhanced by the right kind of *clustering*—bringing resources into close proximity with one another. But it's a special kind of clustering, one whose hallmark is density and diversity—of talent, resources, tools, models, and ideas. It's this kind of clustered diversity that ensures the resilience of innovation hubs like Silicon Valley and the old-growth forest alike.

These principles—tight feedback loops, dynamic reorganization, built-in countermechanisms, decoupling, diversity, modularity, simplicity, swarming, and clustering—form a significant part of the tool kit for systemic resilience. Taken together, they form a powerful vocabulary for evaluating the resilience or fragility of the big systems like cities, economies, and critical infrastructure that underwrite our contemporary lives. Using this toolkit, we can ask: How can we create more effective feedback loops between our actions and their consequences? How might we decouple ourselves from a scarce underlying resource, or make our infrastructure more modular?

Understanding these principles also suggests resilience's distinction from, and relationship to, some important related ideas. For example, though the words are often used interchangeably, resilience is not *robustness*, which is typically achieved by hardening the assets of a system. The Pyramids of Egypt, for example, are remarkably robust structures; they will persist for many thousands of years to come, but knock them over and they won't put themselves back together.

The same holds true for *redundancy*, which is also a time-tested way to improve the ability of a system to persist even when compromised, but is also not quite the same thing as resilience. Keeping backups of critical components and subsystems is certainly wise on its face, as anyone who's been stuck on a lonely road with a flat tire and no spare can attest. Highly resilient systems are frequently *also* highly redundant systems. But backups are costly, and in good times there can be a great deal of pressure placed on a system to eliminate them to improve efficiency. Worse still, these backups may become of little or no use when circumstances change dramatically.

Finally, and perhaps most counterintuitively, resilience also does not always equate with the *recovery* of a system to its initial state. While some resilient systems may indeed return to a baseline state after a breach or a radical shift in their environment, they need not necessarily ever do so. In their purest expression, resilient systems may have no baseline to return *to*—they may reconfigure themselves continuously and fluidly to adapt to ever-changing circumstances, while continuing to fulfill their purpose.

None of this is to say that resilient systems never fail. Regular, modest failures are actually *essential* to many forms of resilience—they allow a system to release and then reorganize some of its resources. Moderate forest fires, for example, redistribute nutrients and create opportunities for new growth without destroying the system as a whole. (Paradoxically, they do so by ensuring that fire-resistant species are not crowded out by nonresistant ones as a healthy forest reaches its peak.) When human beings intervene in this cyclical process and prevent these necessary smaller fires from happening, a forest can build up so much kindling that a small accidental fire can become catastrophic. Just ask a Californian.

More broadly, resilient systems fail gracefully—they employ strategies for avoiding dangerous circumstances, detecting intrusions, minimizing and isolating component damage, diversifying the resources they consume, operating in a reduced state if necessary, and self-organizing to heal in the wake of a breach. No such system is ever perfect, indeed just the opposite: A seemingly perfect system is often the most fragile, while a dynamic system, subject to occasional failure, can be the most robust. Resilience is, like life itself, messy, imperfect, and inefficient. But it survives.

FROM SYSTEMS TO PEOPLE

In the latter chapters of this book, we'll turn our attention from the resilience of systems to that of the people and communities who live with them. As we do, some of the same themes reappear, along with some new ones.

We'll start by exploring new insights into the resilience of individual people. And here there is good news: New scientific research suggests that personal, psychic resilience is more widespread, improvable, and teachable than previously thought. That's because our resilience is rooted not only in our beliefs and values, in our character, experiences, values, and genes, but critically in our *habits of mind*—habits we can cultivate and change.

As we expand our frame to consider the resilience of groups, new themes emerge. The most important of these is the critical role of *trust and cooperation*—people's ability to collaborate when it counts. We'll look at two cases of cooperation in the midst of a crisis, one from Haiti and one from Wall Street—the former spectacularly successful, the latter spectacularly unsuccessful—and explore concrete things we can do to build, and harness, collaborative systems.

Also, as we'll see again and again, establishing a "warm zone" of *diversity* plays an enormous role in resilience and is one of its most important correlates. Whether it's the biodiversity of a coral reef or, in the social context, the cognitive diversity of a group, increasing the diversity of a system's constituent parts ensures the widest palette of latent, ready responses to disruption. The trick is to balance such diversity with mechanisms that ensure that these diverse actors can still cooperate with one another when circumstances dictate.

In our travels, wherever we found strong social resilience, we also found strong communities. And here we don't mean wealthy. Resilience is not solely a function of the community's resources (though of course those help) nor defined solely by the strength of their formal institutions (ditto). Instead, we found resilient communities frequently relied as much on *informal* networks, rooted in deep trust, to contend with and heal disruption. Efforts undertaken to impose resilience from above often fail, but when those same efforts are embedded authentically in the relationships that mediate people's everyday lives, resilience can flourish.

Finally, when we found a resilient community or organization, we almost always also found a very particular species of leader at or near its core. Whether old or young, male or female, these *translational leaders* play a critical role, frequently behind the scenes, connecting constituencies, and weaving various networks, perspectives, knowledge systems, and agendas into a coherent whole. In the process, these leaders promote adaptive governance—the ability of a constellation of formal institutions and informal networks to collaborate in response to a crisis.

These elements—beliefs, values, and habits of mind; trust and cooperation; cognitive diversity; strong communities, translational leadership, and adaptive governance—make up the rich soil in which social resilience grows. Taken together, they suggest new ways to bolster the resilience of communities and organizations, and the people who live within them.

The concept of resilience is a powerful lens through which we can view major issues afresh: from business planning (how do we hedge our corporate strategy to deal with unforeseen circumstances?) to social development (how do we improve the resilience of a community at risk?) to urban planning (how do we ensure the continuity of urban services in the face of a disaster?) to national energy security (how do we achieve the right mix of energy sources and infrastructure to contend with inevitable shocks to the system?). These all affect the one circumstance that matters to each of us: our own (how do we ensure our personal resilience in the face of life's inevitable hardships?).

In all of these contexts, resilience forces us to take the possibility—even necessity—of failure seriously, and to accept the limits of human knowledge and foresight. It assumes we don't have all the answers, that we'll be surprised, and that we'll make mistakes. And while we advocate for it here as a desirable goal, resilience—a property of systems—is not always a virtue in and of itself: Terrorists and criminal organizations are also highly resilient, often for the same reasons listed above. As we'll see, when exploring resilience, we often have as much to learn from the "bad guys" as we do from the "good."

Yet resilience-thinking does not simply call us into a defensive crouch against uncertainty and risk. Instead, by encouraging adaptation, agility, cooperation, connectivity, and diversity, resilience-thinking can bring us to a different way of being in the world, and to a deeper engagement with it. Bolstering our chances of surviving the next shock is important, but it's hardly the sole benefit.

In the discussion that follows, there are other recurring principles. The first of these is *holism*. In a complex system, bolstering the resilience of only one part or level of organization can sometimes (unintentionally) introduce a fragility in another, which in turn can doom the whole. Considering the connected whole, on the other hand, can work to our advantage: When we do so, efforts we undertake in one part of a system can unlock greater resilience in another.

The deeper lesson is that to improve resilience we often need to work in more than one mode, one domain, and one scale at a time—we have to think about the aspects of a system that move both more slowly and more quickly than the one we are interested in, or examine aspects that are, at once, more granular and more global. Consider the constellation of forces at work in the story of the tortilla riots, for example: Some, like Katrina, moved very quickly; others, like the strengthening correlation between corn and oil prices, moved more moderately; and still others, like the economic concentration of international import markets, moved more slowly. It was the shearing forces between these different systems, each with their own velocity, that amplified the disruption. No attempt to correct the system could succeed for long without accounting for their interplay.

You will also notice that strategies for resilience often distill principles that are given their purest expression in actual, living things, whether embodied in an individual cell, a species, or an entire ecosystem. This should hardly be surprising—resilience is a common characteristic of dynamic systems that persist over time, and life on Earth is the most dynamic and persistent system anyone has ever encountered.

Yet this is not an endorsement of some vague "Kumbayah" naturalism. Living systems are messy and complex, and they operate in ways that are less than perfectly efficient—they are in a state of constant, dynamic *dis*equilibrium. Encoded within each one are a diverse array of latent tools and strategies that are only occasionally, if ever, called upon. Carrying around this menagerie of rarely used but useful mechanisms imposes a real cost on the cell, the organism, or the ecosystem by increasing its complexity, slowing its growth, decreasing its peak efficiency, and limiting the resources available to nourish individual components at the expense of the whole.

Similarly, applying such strategies to the real world of human affairs isn't easy, certainly not politically. Doing so involves trading away the certainty of short-term efficiency gains for the mere *possibility* of avoiding or surviving a hypothetical future emergency which may never materialize. Selling umbrellas in the sunshine is a tough job in the best of circumstances (ask any civic or corporate leader), but it's even more difficult in the short-term-obsessed world of impatient stockholders, quarterly earnings reports, biannual elections, and constrained municipal budgets. If it were otherwise, we'd be living in a less bubble-prone world, and no one would groan as they lined up for a fire drill.

Living systems are also profoundly *cyclical*, rooted in what ecologist C. S. "Buzz" Holling, one of the founding figures of resilience research, termed the "adaptive cycle." This is marked by four discrete, looping phases, starting with a rapid *growth* phase in which underlying resources come together, begin to interact, and build on top of one another, like an early-stage forest. This is followed by a *conservation* phase, in which, like a more mature forest, the system becomes increasingly efficient at locking up and utilizing resources, but also becomes increasingly *less* resilient as it does so. This is followed by a *release* phase, in which resources are dispersed, often in response to a disruption or collapse, and then finally a *reorganization*, when the cycle begins anew.

While not every system goes through precisely the same process, the adaptive cycle helps us understand the resilience of many entities beyond the realm of ecology. In industry, for instance, the adaptive cycle is omnipresent. Think how often this story is repeated (and experienced) in business: An entrepreneurial company creates a new and highly desirable product or service. By optimizing that innovation and rigorously eliminating alternatives, it grows very rapidly. It becomes a highly profitable incumbent, squeezing out smaller rivals. Then, when a competitor's disruptive innovation takes hold, it suddenly finds that the very optimization that made it successful is now maladaptive, and the organization rapidly declines. As a result, human resources are released to start the entrepreneurial cycle anew. Growth, conservation, release, reorganization. Sound familiar? It's the story of the rise and fall of Detroit in the face of the oil shocks of the 1970s; the story of Microsoft's rise and fall in the face of the web in the 1990s; the story of Sony and the iPod in the 2000s.

A related theme in the resilience discussion is the importance of networks, which provide a universal, abstract reference system for describing how information, resources, and behaviors flow through many complex systems. Having a common means to describe biological, economic, and ecological systems, for example, allows researchers to make comparisons between the ways these very different kinds of entities approach similar problems, such as stopping a contagion—whether an actual virus, a financial panic, an unwanted behavior, or an environmental contaminant—when it begins to spread. Having a shared frame of reference allows us to consider how successful tactics in one domain might be applied in another—as we'll see in newly emerging fields like ecological finance.

Of course, most of the pressing challenges that confront us exist at a different kind of boundary: the one where people and some other technical, ecological, financial, or social system interact. In the language of systems analysis, these human and nonhuman systems are *coupled*— the behavior of each system influences the other in complex feedback loops whose effects may be difficult to trace—just as the price of corn was coupled to the price of oil after Hurricane Katrina. Unfortunately, as we'll see, it's the tendency of most coupled systems to become brittle over time—to lose rather than gain in their ability to adapt. When that happens, a systemic flip—frequently to a less desirable state of affairs— becomes more difficult to avoid.

Globalization is, in some sense, the mother of all coupled systems, and for all its benefits and wonders, it has often accelerated the loss of adaptive capacity by spinning incomprehensibly vast and interconnected webs across the planet, increasing the latent dependencies between far-flung entities of all types. Globalization has often allowed us to optimize a single variable—for example, resource extraction or consumption—

and temporarily delay or hide the environmental feedback associated with that optimization. Globalization also binds together systems with radically different time signatures—financial transactions that happen in milliseconds, social norms that evolve over years, and ecological processes that normally take millennia. As these interactions grow, the possible sources, speed, and consequences of disruption are all magnified, as is the pain we feel—in our individual lives, communities, institutions, and environment—when that disruption arrives.

MITIGATION, ADAPTATION, AND TRANSFORMATION

This ever-increasing complexity and fragility has spurred a whole spectrum of social and political responses. Some, in what we might call the "Icarus" camp, tell us that we need to prune humanity's footprint, slow down, simplify, and think local. British and American activists in this camp, for example, are already planning so-called Transition Towns—communities that are designed to weather an abrupt anticipated end of global oil supplies and the equally abrupt arrival of climate change. Their strategies are designed to lessen a community's dependence on the larger, hydrocarbon-dependent economy, using means ranging from backyard farming to local energy production. To many in the Icarus camp, an imminent collapse isn't something to be feared, but rather something to be embraced, since it will bring about a more balanced, less consumptive, and more rewarding way of life.

While the followers of Icarus promote a return to smaller scales, members of a "Manifest Destiny" camp argue that it's impossible to turn back, and that we will have to engineer our way out of inevitable problems. For good or ill, they argue, we humans run the planet; with billions of wealthy, wasteful people walking the earth, billions of poor people striving for the opportunity to join them, and billions more as yet unborn about to swell humanity's ranks, the exploitation of resources is unpreventable.

While acknowledging this is troubling, members of the Manifest Destiny camp suggest that these challenges will also provide the spur for new, ever more efficient innovations, many of which are already becoming available, and which can lead us ultimately into something approaching equilibrium with our terrestrial home. In the meantime, what we must do is accept responsibility for and alleviate our inevitable impact on the planet by using technological tools already at hand. Since we can't be smaller, we must be smarter.

Between these poles, much of the conversation in the last decade about adapting to global risks has been shaped by a now more commonplace framework, that of sustainability.

As originally envisioned, the goal of sustainability—to achieve a broad equilibrium between humanity and our planet—was both admirable and inarguable. But as a practical organizing principle, sustainability is now looking increasingly long in the tooth. In part, this is natural: Most ideas have a social life, and a half-life, and at four decades old, the sustainability movement has lasted much longer than other movements. In that time, however, what counts as "sustainable" has been continuously expanded to the point of meaninglessness. (Our favorite recent example: Del Monte Produce pitched its plastic-wrapped, unpeeled single-serve bananas as "sustainable" because they would keep fresher longer in vending machines, necessitating fewer deliveries. If covering an unpeeled fruit in petroleum-based packaging is sustainable, what isn't?)

More seriously, sustainability suffers in two respects: First, the entire notion that the goal should be to find a single equilibrium point runs counter to the way many natural systems actually work—the goal ought to be healthy dynamism, not a dipped-in-amber stasis. Second, sustainability offers few practical prescriptions for contending with disruptions precisely at the moment we're experiencing more and more of them. Resilience-thinking, on the other hand, can provide a broader, more dynamic, and more relevant set of ideas, tools, and approaches. As volatility continues to hold sway, resilience-thinking may soon come to augment or supplant the sustainability regime altogether.

To see why, consider this thought experiment:

Imagine we gather all of the people who are concerned about a major global disruption—irrevocable climate change, for example, though it could be any major future crisis—and place them, metaphorically, in a single car. (For the moment, let's leave out the folks who don't believe in climate change, or don't believe it's a big deal.) Now (in the experiment only!) let's send that car accelerating toward a cliff—a climatological point of no return.

At the beginning of the car's journey, one group in the car will hold moral authority: those who align themselves with risk *mitigation*. "Turn back!" they shout. "Hit the brakes! Or at the very least take your foot off the accelerator!" At this point in the car's journey, this is precisely the moral and proper thing to do.

However, as their calls go unheeded, and the car approaches a point where, even if the brakes were hit, the car would still likely skid over the edge, another group will come to occupy the moral high ground: those who align themselves with risk *adaptation*. "We had better build some air bags and a parachute," they say, "since we could go over whether we like it or not." As above, at this point in the car's journey, this is a moral and proper thing to advocate.

In between these two points, there is usually a transition—sometimes generational—between those who believe the best path is to avoid the danger and those who want to prepare for its aftermath. Early on, the mitigationists accuse the adaptationists of throwing in the towel and conceding defeat too early; later, the adaptationists accuse the mitigationists of wasting time and diverting resources trying to stop the inevitable.

Broadly speaking, the contemporary sustainability movement has been (rightfully) preoccupied with risk mitigation for some time. Yet as irrevocable global changes of all sorts edge closer, a shift toward adaption—and with it, an increasing focus on resilience—is under way. And not just in sustainability, but in many areas of significant future risk—from global economics to public health, poverty alleviation to corporate strategy.

This is not to say that we must abandon hope and accept every calamity as inevitable. Rather, the resilience frame suggests a different, complementary effort to mitigation: to redesign our institutions, embolden our communities, encourage innovation and experimentation, and support our people in ways that will help them be prepared and cope with surprises and disruptions, even as we work to fend them off. This in turn buys us time to embrace longer-term transformation— what we, following the preceding metaphor, might think of as the wholesale reinvention of both the car in which we are riding and the cliff it is approaching. Give the car wings, and you change the context so utterly that we would eliminate the need for brakes or parachutes altogether.

But in order to do any of that, we must first understand where fragilities come from. So that is where we will turn first.

1

ROBUST, YET FRAGILE

This is a book about why things bounce back.

In the right circumstances, all kinds of things do: people and communities, businesses and institutions, economies and ecosystems. The place in which you live, the company at which you work, and even, without realizing it, you yourself: Each is resilient, or not, in its own way. And each, in its own way, illuminates one corner of a common reservoir of shared themes and principles. By understanding and embracing these patterns of resilience, we can create a sturdier world, and sturdier selves to live in it.

But before we can understand how things knit themselves back together, we have to understand why they fall apart in the first place. So let's start with a thought experiment:

Imagine, for a moment, that you are a tree farmer, planting a new crop on a vacant plot of earth. Like all farmers, to get the most use out of your land, you will have to contend with any number of foreseeable difficulties: bad weather, drought, changing prices for your product, and, perhaps especially, forest fires.

Fortunately, there are some steps you can take to protect your tree farm against these potential dangers. To lower the risk of fire, for example, you can start by planting your saplings at wide but regular intervals, perhaps 10 meters apart. Then, when your trees mature, none of their canopies will touch, and a spark will have trouble spreading from tree to tree. This is a fairly safe but fairly inefficient planting strategy: Your whole crop will be less likely to burn to the ground, but you will hardly be getting the most productive use out of your land.

Now imagine that, to increase the yield of your farm, you randomly start planting saplings in between this grid of regularly spaced trees. By definition, these new additions will be much closer to their neighbors, and their canopies will grow to touch those of the trees surrounding them. This will increase your yield, but also your risk: If a fire ignites one of these randomly inserted trees, or an adjoining one, it will have a much higher chance of spreading through its leaves and branches to the connected cluster of trees around it.

At a certain point, continuing to randomly insert trees into the grid will result in all of the trees' canopies touching in a dense megacluster. (This is often achieved when about 60 percent of the available space is filled in.) This, in contrast to your initial strategy, is a tremendously efficient design, at least from a planting and land-use perspective, but it's also very risky: A small spark might have disastrous consequences for the entire crop.

Of course, being a sophisticated arborist, you're unlikely to plant your trees in either a sparse grid or a dense megacluster. Instead, you choose to plant groves of densely planted trees interspersed with *roads*, which not only provide you access to the interior of your property but act as firebreaks, separating one part of the crop from another and insuring the whole from a fire in one of its parts.

These roads are not free: By reducing the area usable for planting, each imposes a cost, so you must be careful where you put them; adding

too many roads is just as bad as adding too few. But eventually, with lots of trial and error, and taking into account local variations in the weather, soil, and geography, you may alight upon a near-perfect design for your particular plot—one that maximizes the density of trees while making smart and efficient use of roads to access them. Your tree farm's exceptional design will easily withstand the occasional fire, without ever burning entirely to the ground, all the while providing you with seasonally variable but not wildly gyrating timber returns.

Imagine your horror, then, when one day you discover that much of your perfectly designed plot has been infested by an invasive foreign beetle. This tiny pest, native to another geographic region entirely, stowed away on a shipment from an overseas supplier, then hitchhiked its way to the heart of your tree farm by clinging to your boot. Once inside, it exploited the very features of your design—your carefully placed roads—that were intended to insure against the risks you thought you'd confront.

It is at this moment in our thought experiment that you have painfully discovered that, in systems terms, your tree farm design is *robust-yet-fragile* (or RYF), a term coined by California Institute of Technology research scientist John Doyle to describe complex systems that are resilient in the face of anticipated dangers (in this case forest fires) but highly susceptible to unanticipated threats (in this case exotic beetles).

On any given day, our news media is filled with real-world versions of this story. Many of the world's most critical systems—from coral reefs and communities to businesses and financial markets—have similar robust-yet-fragile dynamics; they're able to deal capably with a range of normal disruptions but fail spectacularly in the face of rare, unanticipated ones.

As in our tree-planting example, all RYF systems involve critical trade-offs, between efficiency and fragility on the one hand and inefficiency and robustness on the other. A perfectly efficient system, like the densely planted tree farm, is also the most susceptible to calamity; a

perfectly robust system, like the sparsely planted tree farm, is too inefficient to be useful. Through countless iterations in their design (whether
the designer of a system is a human being or the relentless process of
natural selection), RYF systems eventually find a midpoint between
these two extremes—an equilibrium that, like our roads-and-groves
tree farm design, balances the efficiency and robustness trade-offs particular to the given circumstance.

The complexity of the resulting compensatory system—the network of the roads and groves in the example above—is a by-product of
that balancing act. Paradoxically, over time, as the complexity of that
compensatory system grows, it becomes a source of fragility itself—
approaching a tipping point where even a small disturbance, if it occurs
in the right place, can bring the system to its knees. No RYF design can
therefore ever be "perfect," because each robustness strategy pursued
creates a mirror-image (albeit rare) fragility. In an RYF system, the
possibility of "black swans"—low-probability but high-impact events—
is engineered in.

The Internet presents a perfect example of this robust-yet-fragile
dynamic in action. From its inception as a U.S. military funded project
in the 1960s, the Internet was designed to solve a particular problem
above all else: to ensure the continuity of communications in the face
of disaster. Military leaders at the time were concerned that a preemp-
tive nuclear attack by the Soviets on U.S. telecommunications hubs
could disrupt the chain of command—and that their own counterstrike
orders might never make it from their command bunkers to their intended recipients in the missile silos of North Dakota. So they asked
the Internet's original engineers to design a system that could sense and
automatically divert traffic around the inevitable equipment failures
that would accompany any such attack.

The Internet achieves this feat in a simple yet ingenious way: It
breaks up every email, web page, and video we transmit into packets
of information and forwards them through a labyrinthine network of
routers—specialized network computers that are typically redundantly

connected to more than one other node on the network. Each router contains a regularly updated routing table, similar to a local train schedule. When a packet of data arrives at a router, this table is consulted and the packet is forwarded in the general direction of its destination. If the best pathway is blocked, congested, or damaged, the routing table is updated accordingly and the packet is diverted along an alternative pathway, where it will meet the next router in its journey, and the process will repeat. A packet containing a typical web search may traverse dozens of Internet routers and links—and be diverted away from multiple congestion points or offline computers—on the seemingly instantaneous trip between your computer and your favorite website.

The highly distributed nature of the routing system ensures that if a malicious hacker were to disrupt a single, randomly chosen computer on the Internet, or even physically blow it up, the network itself would be unlikely to be affected. The routing tables of nearby routers would simply be updated and would send network traffic around the damaged machine. In this way, it's designed to be robust in the face of the anticipated threat of equipment failure.

However, the modern Internet is extremely vulnerable to a form of attack that was unanticipated when it was first invented: the malicious exploitation of the network's open architecture—not to route around damage, but to engorge it with extra, useless information. This is what Internet spammers, computer worms and viruses, botnets, and distributed denial of service attacks do: They flood the network with empty packets of information, often from multiple sources at once. These deluges hijack otherwise beneficial features of the network to congest the system and bring a particular computer, central hub, or even the whole network to a standstill.

These strategies were perfectly illustrated in late 2010, when the secrets-revealing organization WikiLeaks began to divulge its trove of secret U.S. State Department cables. To protect the organization from anticipated retaliation by the U.S. government, WikiLeaks and its supporters made copies of its matériel—in the form of an encrypted insurance file containing possibly more damaging information—available on thousands of servers across the network. This was far more than the United States could possibly interdict, even if it had had the technical capability and legal authority to do so (which it didn't). Meanwhile, an unaffiliated, loose-knit band of WikiLeaks supporters calling itself Anonymous initiated distributed denial-of-service attacks on the websites of companies that had cut ties with WikiLeaks, briefly knocking the sites of companies like PayPal and MasterCard off line in coordinated cyberprotests.

Both organizations, WikiLeaks and Anonymous, harnessed aspects of the Internet—redundancy and openness—that had once protected the network from the (now-archaic) danger that had motivated its invention in the first place: the threat of a strike by Soviet missiles. Four decades later, their unconventional attacks (at least from the U.S. government's perspective) utilized the very features of the network originally designed to prevent a more conventional one. In the process, the attacking organizations had proven themselves highly resilient: To take down WikiLeaks and stop the assault of Anonymous, the U.S. government would have had to take down the Internet itself, an impossible task.

Doyle points out a dynamic quite similar to this one at work in the human immune system. "Think of the illnesses that plague contemporary human beings: obesity, diabetes, cancer, and autoimmune diseases. These illnesses are malignant expressions of critical control processes of the human body—things like fat accumulation, modulation of our insulin resistance, tissue regeneration, and inflammation that are so basic that most of the time we don't even think about them. These control processes evolved to serve our hunter-gatherer ancestors, who had to store energy between long periods without eating and had to maintain their glucose levels in their brain while fueling their muscles. Such biological processes conferred great robustness on them, but in today's era of high-calorie, junk-food-saturated diets, these very same essential systems are hijacked to promote illness and decay."

To confront the threat of hijacking, Internet engineers add

sophisticated software filters to their routers, which analyze incoming and outgoing packets for telltale signs of malevolent intent. Corporations and individuals install firewall and antivirus software at every level of the network's organization, from the centralized backbones right down to personal laptops. Internet service providers add vast additional capacity to ensure that the network continues to function in the face of such onslaughts.

This collective effort to suffuse the system with distributed intelligence and redundancy may succeed, to varying degrees, at keeping some of these anticipated threats at bay. Yet, even with such actions, potential fragility has not been eliminated; it's simply been moved, to sprout again from another unforeseeable future direction. Worse, like all RYF systems, over time the complexity of these compensatory systems—antivirus software, firewalls—swell until they become a source of potential fragility themselves, as anyone who's ever had an important email accidentally caught in a spam filter knows all too well.

Along the way, paradoxically, the very fact that a robust-yet-fragile system continues to handle commonplace disturbances successfully will often mask an intrinsic fragility at its core, until—surprise!—a tipping point is catastrophically crossed. In the run-up to such an event, everything appears fine, with the system capably absorbing even severe but anticipated disruptions as it was intended to do. The very fact that the system continues to perform in this way conveys a sense of safety. The Internet, for example, continues to function in the face of inevitable equipment failures; our bodies metabolize yet another fast-food meal without going into insulin shock; businesses deal with intermittent booms and busts; the global economy handles shocks of various kinds. And then the critical threshold is breached, often by a stimulus that is itself rather modest, and all hell breaks loose.

When such failures arrive, many people are shocked to discover that these vastly consequential systems have no fallback mechanisms, say, for resolving the bankruptcy of a major financial institution or for the capping of a deep-sea spill.

And in the wake of these catastrophes, we end up resorting to simplified, moralistic narratives, featuring cartoon-like villains, to explain why they happened. In reality, such failures are more often the result of the almost imperceptible accretion of a thousand small, highly distributed decisions—each so narrow in scope as to seem innocuous—that slowly erode a system's buffer zones and adaptive capacity. A safety auditor cuts one of his targets a break and looks the other way; a politician pressures a regulator on behalf of a constituent for the reduction of a fine; a manager looks to bolster her output by pushing her team to work a couple of extra shifts; a corporate leader decides to put off necessary investments for the future to make the quarterly numbers.

None of these actors is aware of the aggregate impact of his or her choices, as the margin of error imperceptibly narrows and the system they are working within inexorably becomes more brittle. Each, with an imperfect understanding of the whole, is acting rationally, responding to strong social incentives to serve a friend, a constituent, a shareholder in ways that have a significant individual benefit and a low systemic risk. But over time, their decisions slowly change the cultural norms of the system. The lack of consequences stemming from unsafe choices makes higher-risk choices and behaviors seem acceptable. What was the once-in-a-while exception becomes routine. Those who argue on behalf of the older way of doing things are perceived to be fools, paranoids, or party poopers, hopelessly out of touch with the new reality, or worse, enemies of growth who must be silenced. The system as a whole edges silently closer to possible catastrophe, displaying what systems scientists refer to as "self-organized criticality"—moving closer to a critical threshold.

This dynamic—and our first hints about what we might do to improve the resilience of such systems—can be seen in two very different robust-yet-fragile systems that, with new analytical tools, are beginning to powerfully illuminate each other: coral reef ecology and global finance.

OF FISHERIES AND FINANCE

In the 1950s, Jamaica's coral reefs were a thriving picture-postcard example of a Caribbean reef, supporting abundant mixtures of brilliantly colored sponges and feather-shaped octocorals sprouting up from the hard coral base. The reefs were popular habitats for hundreds of varieties of fish, including large predatory species such as sharks, snappers, groupers, and jacks, which were harvested by local fishermen as a reliable and time-honored food source for the island's population.

By the 1970s, things appeared relatively unchanged. In the intervening two decades, however, Jamaica's population had swelled by a third. Local fishermen, struggling to feed the island's growing population, had begun using motorized canoes to harvest not only the predator species, but the smaller, plant-eating fish, such as surgeonfish and parrotfish, as well. The reefs still appeared healthy, however, and most of their habitants seemed to be thriving. Sea urchins, in particular, were flourishing since they no longer had to compete with the plant-eating fish for algae, the mainstay of their diet.

Then, on August 6, 1980, after almost four decades without a major storm, one of the most powerful Caribbean tempests in history, Hurricane Allen, descended on the island and its surrounding reefs. Winds exceeding 175 miles per hour whipped up a forty-foot-high storm surge, which pounded the reefs. Shallow-water corals were devastated; however, their deeper-waters cousins, well below the surface, emerged relatively unscathed. In fact, a few months after the hurricane's strike, substantial coral recruitment—the measure of young corals entering the adult population—was found in the deeper waters. For the next three years, coral cover slowly increased.

The general consensus among marine biologists at the time was that Jamaica's reefs had survived remarkably well after the battering of Hurricane Allen. Data seemed to suggest that the reef system was surviving and, in the deeper waters around the island at least, maybe even thriving.

Then, in 1983, something terrifying happened below the surface of Jamaica's waters. An unidentified pathogen decimated the entire population of Jamaica's long-spined sea urchins. The illness was unprecedented in its lethality and its speed: Within a few days of the onset of symptoms, one observer reported, "on reefs that used to be black with urchins, one could swim for an hour without seeing a single living individual." By February 1984, the sea urchin had been virtually eliminated from its normal habitat—the most extensive and severe mass mortality ever reported for a marine organism.

Coming after the long and slow decline of the other native plant-eating fish species from overfishing, the loss of the urchins proved catastrophic to Jamaica's reefs. With no urchins—or other species—to keep it in check, algae quickly came to dominate every corner of the reef system, eventually covering 92 percent of its surface area and killing the corals underneath. With the loss of the corals went the remaining fish—reefs that supported hundreds of species for thousands of years were transformed into vacant algal wastelands seemingly overnight.

On a healthy reef, a new pathogen decimating a *single species* (like the urchin) might not have had catastrophic consequences, because an essential reef function—like keeping algae in check—could be performed by more than one species. On the highly compromised Jamaican reef, however, the continued flourishing of the ecosystem as a whole became entirely dependent on a single species continuing to do that job. The loss of the urchins, an otherwise modest trigger, caused the reef to collapse virtually overnight.

Yet, had you asked a marine scientist in 1982 to describe the reefs, you would have received a promising assessment: The reef had proved robust in the face of significant disturbances, ranging from hurricanes to extensive human fishing. There was little evidence of the hidden fragility that was being exacerbated by the slow loss of biodiversity.

Such clarity is available to us in hindsight, but consider the challenges posed by trying to manage such a system at the time. The interactions between the various agents affecting the health of the reef (fish, urchins, algae, corals, and human beings) were imperfectly understood and nonlinear—small changes could lead to big outcomes, and vice versa. Some of the dependencies between these agents were masked: Prior to the loss of the urchins, it was hard to tell if the loss of herbivorous fish was having a significant impact or no impact whatsoever. What's more, recent experience suggested that the system could rebound from a disturbance on the scale of a hurricane. And even a healthy reef can behave in highly variable ways. Fish stocks rise and fall—how would you separate normal variability from the onset of collapse?

Similar questions confront us whenever we try to manage a complex system with highly interdependent parts. Whether we're dealing with fish stocks or financial stocks, to improve the resilience of the system as a whole, we first need measurement tools that take the health of whole systems into account, not just their pieces. At least if we want to keep eating seafood.

As it happens, in the 1950s, the same decade that our story in Jamaica began, the California sardine industry was also devastated by a collapse. The booming fishing industry had provided the dramatic backdrop for John Steinbeck's novel *Cannery Row*: In the mid-1930s, 790,000 tons of sardines were commercially fished from California's waters. Yet by 1953, the catch had plummeted to less than 15,000 tons—a drop of 98 percent.

Two competing hypotheses emerged to explain the collapse: one based on traditional simple overfishing and another on the La Niña currents that cooled California's ocean waters. However, in 2006, after poring over fifty years of data on sardine larvae, George Sugihara, a mathematician and theoretical ecologist at the Scripps Institution of Oceanography, proved that both theories were wrong. The sardine stock had collapsed not because the fishing industry was taking out too many *small* fish, but because it was taking out too many *big* ones: Sugihara found that California's industrialized fishing operations had become so efficient at catching adult sardines that they had significantly changed the age structure of the entire stock. Bereft of adults in 1949 and 1950, the substantially more juvenile sardine population had failed to spawn, and when it encountered additional stressors from the natural world, it flipped into collapse.

Given sufficient time, it's possible that the sardine population could have recovered from this megabust all by itself—such instabilities do occasionally occur naturally in ecosystems from time to time. Unfortunately, fisheries management practices of the time rarely took these kinds of instabilities into account. For much of the twentieth century (and in many places around the world today) the status quo of most fisheries management was based on maximum sustainable yield (MSY), or the largest catch that can be fished from the stock of a species over an indefinite period of time.

MSY is predicated on a linear system with stable equilibrium points. By addressing only one species at a time, and assuming that all other variables were fixed and unchanging, the MSY regime didn't call for moderation from California's commercial fishing fleets when sardine stocks started to decline. Rather, in the face of the intensified economic pressure of a reduced catch, many local fishing operations accelerated their harvesting efforts, pushing the ecosystem past the point of recovery. The fishing system had no breaks.

And MSY is still in widespread use today. It has been implicated in declining fish stocks all over the world: Today 63 percent of all fished species are overfished and in danger, and 29 percent have collapsed outright, meaning they are now, like the sardine, at least 90 percent below their historic maximum catch levels. In 2006, an international research team led by Boris Worm of Dalhousie University in Canada calculated that, if these trends continue, by 2048 all commercial fishing on Earth will cease—there will simply be no more fish in the sea.

To prevent such a catastrophe, Sugihara and others are trying to promote a more holistic alternative to MSY called ecosystem-based fishery management (EBFM). This whole-systems model is rooted in the observation that ecosystems are difficult to model and predict, with constantly changing threshold points that, when exceeded—as in the case of the California sardine population—can cause the system to collapse or restructure. EBFM counteracts this by promoting the maintenance of biodiversity as a central management goal whenever and wherever possible, at every level of organization, from small niches to entire regions of the sea.

Doing so starts with a very different system for measuring what's going on in an ecosystem that's being fished. Under MSY, fisheries managers gathered and analyzed information primarily about the species in the catch and little else; EBFM, on the other hand, requires them to measure the species of fish that are *not* caught as well as the ones that are. Many other factors, referred to as *ecosystem indicators*, need to be gathered and correlated as well, such as coastal upwelling, the pattern of wind and water movement that brings cool, nutrient-rich waters to ocean's surface near the coast, where they provide the diet for plankton, which in turn form the base of the food chain. Critically, EBFM also calls for measurement and correlation of societal trends on land as well as what's happening in the water—merging the social and ecological pictures.

Such a holistic management regime, based on nonlinear, complex systems, has gained traction politically, but implementing EBFM remains challenging: Very few of the fisheries management institutions have the expertise, history, resources, and know-how to turn away from the more straightforward equilibrium models that underpin the maximum sustainable yield regime.

"We like to see the world as consisting of separate parts that can be studied in an isolated, linear way, one piece at a time," Sugihara explains. "These pieces can then be summed independently to make the whole. Researchers have developed a very large tool kit of analytical methods and statistics to do just this, and it has proven invaluable for studying simple engineered devices. But we persist with these linear tools and models even when systems that interest us are complex and nonlinear. It is a case of looking under the lamppost because the light is better even though we know the keys are lost in the shadows."

To get fisheries to switch, ecologists are now working to make EBFM principles easier to implement, and to do so they're borrowing a central concept from finance: the portfolio model. Interacting species in the ecosystem are linked together as a set of correlated assets, and, just as in a financial portfolio, fisheries managers make decisions that balance risks and returns among them. The portfolio method suggests that a particular species should be assessed on the basis of how it contributes to the overall performance of the ecosystem, rather than in isolation—in much the same way that a particular financial stock contributes to the value of a diversified fund. In both cases, there is no intrinsic value to the asset or the species—context is everything.

Managing entire classes of interacting species in this way simplifies what are otherwise very complex and nonlinear dependencies. By focusing on the relationships among categories of species, and not their absolute numbers, a portfolio approach can more easily accommodate complex factors like environmental fluctuations and advances in fishing technologies, even as it makes potential risks to the system much more explicit. And a multispecies portfolio can also be recalibrated according to the changing conditions of the local ecosystem, much like a financial adviser might arrange a portfolio one way for a young, risk-tolerant active investor and very differently when that investor becomes a risk-averse retiree.

The portfolio approach has other benefits as well. In a financial market, investing in groups of assets allows investors to lower the variability of their returns, in exchange for increasing the reliability of those returns. A properly balanced portfolio incorporates various strategies and hedges to smooth out the troughs and the peaks: During boom times, having commodities mixed in with aggressive growth stocks might keep you from generating sufficient returns to buy a Ferrari, but during a bust it will also keep you out of the poorhouse. Similarly, maintaining a balanced "portfolio" of species allows a fisheries manager to reduce the variability of the annual catch—a major benefit to both fish and fishermen—in exchange for greater stability. In one study of the Chesapeake Bay, ecological economists Martin Smith and Douglas Lipton found that, using just such a portfolio-driven method, over the forty-year period from 1962 to 2003, the bay's fisheries managers could have generated better financial returns from fishing and reduced the season-to-season variability of the catch—a true win-win.

In matters of finance and fisheries, portfolio-driven approaches provide us with choices: If we seek a specific level of return, there are portfolios that can be designed to deliver that return while minimizing the risks; conversely, if we are comfortable with a particular level of risk, there are portfolios that can be designed at that level of risk to maximize our return.

ECOLOGY FOR BANKERS

Even as concepts from financial management begin to inform new strategies for improving the resilience of ecological systems like fisheries, the reverse is also occurring. A new generation of economists and finance experts are beginning to unearth important lessons in ecology for improving the resilience of the global financial economy. These insights are fueling the birth of an entirely new field, called *ecofinance.*

Consider, for a moment, some parallels between the Jamaican coral collapse mentioned earlier and the recent global financial crisis. Just as with the die-off of the sea urchins, the trigger of the financial collapse—Lehman Brothers' $600 billion filing for Chapter 11 bankruptcy in mid-September 2008—was, in the context of a $70 trillion annual global economy, a relatively modest event. Much like the Jamaican reefs' prior survival of Hurricane Allen, the financial markets had robustly weathered a prior decade filled with significant disturbances, including a dot-com bubble, oil shocks, and wars in the Middle East. Yet the decision to let Lehman fail, which we explore in greater detail later in this book, brought about an epidemic of fear and uncertainty that ground the world's entire capital markets to a halt.

Just like the decades-long erosion of biodiversity on the Jamaican reef, the underlying reasons for the financial collapse also involved a number of creeping structural changes to the market that set the stage for catastrophic collapse. And, as we've seen in other RYF systems, like those in the human body or the Internet, when that fragility finally did appear, its effects were amplified by the hijacking of normally beneficial aspects of the global financial system, including its structure, its approach to risk management, its feedback mechanisms, its levels of transparency, and its product innovations.

In 2008, George Sugihara, the same ecologist who had unearthed the reasons for the California sardine collapse, and two other ecologists, Simon Levin and Robert May, published an essay in *Nature* titled "Ecology for Bankers." They wanted to offer guidance for applying the tenets of holistic, ecosystem-based management to the financial markets. Like a densely planted tree farm, both systems had been managed for increased efficiency at the cost of increasing complexity and fragility. And both had implemented similar risk-management techniques: In the same way that maximum sustainable yield practices were applied to lone species of fish, it was common in the banking sector to apply single-firm risk analysis, looking only at individual banks without ever examining interconnection in the system as a whole. Worse, financial firms tended (and still tend) to calculate risk additively, meaning they take individual measures of risk for each transaction they make and add them up to arrive at a picture of their total potential exposure. This type of modeling makes the financial system seem much more secure than it really is; in fact, just as in marine ecosystems, the nonlinear nature of the financial network is multiplicative: Certain failures multiply the risks of other failures.

"Economics is not typically thought of as a global systems problem," Sugihara explains. "Investment banks are famous for a brand of tunnel vision that focuses risk management at the individual firm level and ignores the difficult and costlier systemic perspective. Monitoring the ecosystem-like network of firms with interlocking balance sheets is not in a corporate risk manager's job description. But ignoring these counterparty obligations and mutual interdependencies is exactly what prevented financial institutions from seeing and appropriately pricing risk, which in turn dramatically amplified the recent financial crisis."

The essay in *Nature* came on the heels of an earlier, high-level conference in 2006 sponsored by the Federal Reserve Bank of New York, the U.S. National Academies, and the National Research Council to stimulate fresh thinking on systemic risk. Sugihara and Levin introduced the bankers in attendance to the concept of a *trophic web*—the way in which energy and nutrient flows in an ecosystem connect different species.

The basic concept will be familiar from grade-school science class: Aquatic plants convert the energy of the sun into food; the plants are eaten by small fish; the small fish by bigger fish. As fish and plants die, their bodies provide nutrients to smaller organisms and are recycled through the system. Trophic webs map these energy-transfer relationships in detail.

A similar framework can be used to analyze the transfer of value across a financial network. Yet, while creating detailed trophic webs is a common activity in ecology, creating similarly detailed flows is less common for large-scale financial networks. To remedy this, in 2006, the Federal Reserve Bank of New York commissioned a study of the topology of the interbank payment flows inside what's called the Fedwire system—the mechanism that U.S. banks use for transferring funds between them. Fedwire acts as the backbone of the U.S. financial system, processing an astounding average daily volume of almost 500,000 interbank payments, with a daily value of approximately $2.4 trillion (in 2010, the last year for which data was available at the time of writing). These payments can be thought of as the equivalents of energy and nutrients in the financial ecosystem being transferred between different species of institutions.

The sample used for the study involved approximately 700,000 transfers, and more than five thousand banks, taken from a typical day. The picture that emerges is startling: While most banks had a small number of connections, a few hubs had thousands. At the core of the network, just sixty-six banks accounted for 75 percent of the daily value of transfers. Even more telling, the network topology revealed that twenty-five of the biggest banks were completely connected—so inter-twined that a failure among any strongly suggested a failure for all, the very definition of "too big to fail."

What's true within the core of the U.S. financial sector is increasingly true at an international level as well. An analysis of the connections between eighteen national financial markets reveals that, over the last two decades, central finance hubs around the world, like London and Hong Kong, have swelled to roughly fourteen times their earlier size; at the same time, the links between them have increased sixfold.

The increased size, connectivity, and volume of capital flowing through the financial system are not intrinsically bad—in fact, in healthier circumstances, they ensured that capital could flow where (and when) it was most needed, generating improved returns and keeping the system liquid. This connectivity was celebrated in the precrash era as a tool for dispersing, not concentrating risks.

The market's densely connected configuration, much like the Internet's, ensured that randomly shutting down one of the innumerable banks in the global system would not be likely to cause systemic problems, because, statistically, the vast majority of banks in the network are at the end of spokes connected to a very limited number of hubs. But flip one of those central hubs (a rare and dangerous occurrence), and you might not only take down the thousands of banks directly connected to it, but the other hubs as well, along with thousands of institutions connected to *them*. Like a game of Jenga, pull a random plank and almost certainly nothing will happen. Pull the wrong one, and the entire robust-yet-fragile edifice comes crashing down.

And that's just what happened: For several decades, a string of moderate and even serious failures was capably handled by the financial system, from the dot-com collapse to oil shocks—each one increasing confidence in the ability of the system as a whole to manage disruption. Then a hub failed. Panic spread throughout the network, and things quickly ground to a halt. As Levin says, "It's not that the hubs were too big to fail, but too *interconnected* to be allowed to fail [italics added]." When calamity struck, there was no way to separate or decouple them from one another.

These interconnections in the financial system didn't just take the normal form of funds flowing between banks to cover daily activities. They were anchored by new, sophisticated debt and insurance derivatives: collateralized debt obligations (CDOs) were the securities that allowed banks to cut up, repackage, and then sell one another the debt from risky U.S. home mortgages; credit default swaps (CDSs) were the insurance contracts that tied these banks to one another in massive webs of financial interdependence. Together, they were the financial equivalent of nitric acid and glycerin—in small amounts they keep your heart pumping; in large amounts: *boom*.

FINANCIAL CLUSTER BOMBS

A CDO is a financial instrument that can best be understood by imagining a set of wineglasses stacked in a pyramid. When champagne is poured over the pyramid, the glass at the top is filled first, the glasses in the middle are filled next, and those at the bottom are filled last. A CDO was an equivalent financial instrument, but instead of disbursing wine into wineglasses, it poured the monies from mortgage payments into a set of specialized bonds.

To create a CDO, a bank would bundle together a group of mortgages, held by regular U.S. homeowners. Each month, as these homeowners wrote their monthly mortgage checks, the banks would pool these payments together and make payouts to a series of bonds called *tranches* that were stacked up just like the wineglasses. The tranche at the top of the chain, like the glass at the top of the pyramid, got paid first, then the one following it, and so on, until either the tranche at the bottom was paid or the pool of funds was exhausted.

By definition, the top tranche, at the front of the line to be repaid, was the least risky, so it earned both the best rating (AAA) from ratings agencies like Moody's and Standard & Poor's and also the lowest rate of return, perhaps 2 percent. The bottom tranche, on the other hand, was the most risky: If mortgage holders in the CDO pool stopped paying their mortgages, the loss would be felt in that tranche first—so it earned both the lowest rating (say BB) and the highest rate of return, perhaps 10 percent.

So far, so good. But from there, CDO engineering quickly entered into the twilight zone. A bank might, for example, take the lowest-rated tranche (BB) of a particular CDO (let's call it Lucifer) and turn it into its *own* CDO (let's call this bottom tranche Damien). Even though it was comprised of junk, Damien, through the magic of financial engineering, was divided into its own set of tranches, the top one of which was awarded its own triple-A rating. This absurdity masked the fact that the underlying asset upon which it was based—Lucifer's bottom-most BB tranche—was toxic, high-risk junk. It was like saying "here is the safest house that you can build with this toxic waste" and conveniently forgetting to say the toxic-waste part. Or, if you prefer, like taking a dozen hobbled horses from the glue factory, lining them up according to their speed, and calling the fastest one a Thoroughbred.

When some homeowners originally bundled together under Lucifer stopped being able to pay their mortgages, not only did its BB tranche bondholders lose out, so did those holding *all* of Damien's tranches— even the ones rated AAA. And there's the true tragedy: Those holding Damien's triple-A's were well-managed, risk-averse pension funds, municipalities, and 401(k) plans.

If CDOs were the financial equivalent of chloresterol-laden junk food mislabeled as health food, the credit default swaps were the mechanism by which the banks ensured that a blockage in any artery would give a heart attack to all. CDSs are insurance contracts, much like the contract that you might sign to insure a car, a home, or your own life. In such arrangements, you make a monthly payment to the insurance company, and, in the event of disaster, it makes you whole. Similarly, these insurance contracts enabled a bank to insure against the loss in value of a stock or bond it might purchase. If Bank A buys $10 million in corporate bonds, and it loses $2 million in value, then the insuring Bank B would make up the difference. In the meantime, Bank A would pay a premium on this insurance to Bank B, much as you pay your health and car insurance premiums.

CDSs had several crucial differences from traditional insurance, however. First, a CDS contract could be traded from investor to investor with no oversight or even regulations ensuring that the insurer had the ability to cover the losses when and if it needed to. By calling the contract a swap, and not insurance, the investors in CDSs were able to avoid the capital reserve requirements and regulatory oversight of the traditional insurance industry. (This was presumably the "innovation" at CDSs' core.)

Second, CDSs allow firms not only to insure against the possible default of their own investments, but to insure against the possible default of another company's assets—akin to taking out an insurance policy on your neighbor's Ferrari. Firms could use swaps as a tool for speculation— to bet a company would fail. This practice was made illegal in traditional insurance markets as far back as the 1700s, before which it was legal for individuals to buy insurance on British ships that they didn't own, creating—*quelle surprise*—a rash of perfectly seaworthy ships mysteriously sinking to the bottom of the Thames. In its place, Parliament codified the notion of "insurable interest," the requirement that you have an actual economic interest in the asset being insured. It was a concept that reigned unchallenged for two and half centuries—until the rise of the CDS.

Finally, CDS contracts were sold and traded privately, or "over the counter." While they added enormously to the risk profile of the institution doing the insuring, they didn't show up on the traditional balance sheet. When the crisis came, nobody knew who owed what to whom and what it meant for anyone's bottom line.

In theory, CDOs and CDSs were originally designed to allow the market to do two things that are quite beneficial: first, to distribute risks to those who were most capable and willing to take them, and second, to allow banks to diversify their portfolios by mixing and matching some of their own activities with one another's. A typical big bank might find itself originating its own mortgages, buying them from others, selling mortgage-backed securities that blended the two, and insuring those of another bank against default. All of this looked, superficially at least, like diversification—a wise strategy for balancing efficiency and robustness.

But, in allowing debt, credit, and risk to be sliced into tranches, packaged, bought, repackaged, sold, and resold, these instruments also made the dependencies between institutions mind-bogglingly complicated. The chain of custody for the underlying assets lengthened to the point of incomprehensibility.

Thus, when the crisis hit, none of the banks could be quite sure if the *other* institutions with which they had contracts might *also* be enmeshed in *other* contracts that left those institutions on the hook in some potentially catastrophic way. This is known as the problem of *counterparty risk*—not the risk that you'll become a deadbeat, or the risk that your partners will become deadbeats, but the risk that some of your partners' partners will become deadbeats.

In the low-risk, precrash world, from an individual bank's perspective, having a contract with another bank—which itself had many other business relationships—was perceived as a good thing: It was a measure of diversification and suggested lower risk. After all, what's the likelihood of a significant number of contracts in a large portfolio defaulting all at the same time? In the high-risk, postcrash world, however, having a contract with another bank with lots of counterparties was nightmarish: After all, it might be holding a contract with an unknown third party that might, at any moment, blow up in its face. How could you *possibly* determine how creditworthy the other bank was? How could you (or it) know what its obligations were? Or those of its counterparties' counterparties? How could you trust *anybody?*

When the crash came, a sense of transparency disappeared overnight, and with it went the most important variable in the system, trust—a theme we will revisit later in this book.

It didn't help that the derivative contracts themselves were mind-bogglingly complicated. The simplest of these ran some two hundred pages—the most advanced varieties required reading in excess of one *billion* pages. (Reading one page a minute, it would take you slightly more than 1,900 years to read a single contract for one of these products.) Firms had foregone the due diligence and just swallowed the contracts whole. After the crash, figuring out who owed what to whom wasn't just hard, it was impossible. It's unsurprising that the institutions that vaporized amid the destruction—Lehman, Bear Stearns, and AIG Financial Products—had among the largest counterparty exposure. They were attached to an anchor of unknowable size.

The rise of these complex derivatives had another, subtler impact as well. Their promise—of higher returns and dramatically lower risk—proved so intoxicating to financial organizations of all stripes that these firms effectively homogenized their methods of generating revenue and their risk management strategies to take advantage of them. Many different kinds of actors in the market—from commercial banks to hedge funds—started getting into one another's businesses, holding the same classes of assets and liabilities, in the same proportions, seeking out the same goal, in the same way. As each market player began to internalize more and more of the ever-increasing complexity of the market as a whole, market risks became internal risks. This was diversification without a difference. Right under everyone's noses, the complex ecosystem of the global financial markets was turning into a monoculture—of very complicated lemmings.

THE TELLTALE SIGNS
OF A SYSTEM FLIP

According to Sugihara, in a complex system, there are telltale warning signs of a critical transition, or system flip, and they were visible in
the run-up to the financial crisis. One is a phenomenon called *critical
slowing*—the tendency of a system to become unstable near its threshold point. "When a system is under stress, it can be thrown out of
equilibrium more easily, and it is slower to recover. Without sufficient
recovery time, small perturbations can be amplified until the system is
oscillating wildly out of control—even squealing from one stable state
to another—like a car being oversteered on an icy patch."

Just before such destabilization occurs, a system paradoxically may
experience synchrony, as agents within it briefly behave in lockstep just
before being thrown into chaos. Synchrony can be seen in the brain
cells of epileptics, for example, minutes before the onset of a seizure,
and it was evident in the financial markets prior to the crash. By the
height of the credit boom, from 2004 to 2007, performance across sectors of the financial system was correlated by more than 90 percent, a
reflection of the self-similarity of the various market participants. "This
was clearly an early indicator of impending danger," says Sugihara.

"Synchrony is evident when incentives or pressures lead individual
actors to fall into step and make similar choices," adds Levin. "In un-
synchronized populations, some individuals thrive while others are in
decline; in synchronized populations, a collapse in one place translates
into a collapse in all places."

In a healthy financial system, interconnectivity, risk management,
diversification, and product innovation typically act as shock absorbers,
dispersing risks and mitigating the impact of inevitable failures. But in
the lead-up to the crash, the diversity of the market was slowly eroded,
the actions of the market players became synchronized, and the dependencies between them became unknowable. Then, as with the collapse
of Jamaica's sea urchins, the collapse of Lehman introduced a virus of

unprecedented lethality and speed that decimated the financial system's most critical resource: trust. When that happened, the system's shock absorbers were hijacked and turned into shock amplifiers—spreading the contagion of uncertainty rather than the perception of safety.

In response, the bankers did what seemed rational in such an uncertain situation: They hoarded cash and desperately tried to sell the depressed assets they had on their books. Yet most of the banks had followed similar business strategies before the crash, and so their most rational responses to it were also almost identical. In a densely connected network filled with clones, both responses, taken en masse, made the situation worse for everyone. Hoarding cash caused a liquidity crisis that made it harder for all banks—including the hoarders—to meet day-to-day obligations. The mass sell-off of depressed assets further accelerated the decline in value of the remaining assets on everyone's books, harming the balance sheets of healthier banks and pulling more institutions into the vortex. In this light, it's easy to see why, even with hundreds of billions in bailouts from the U.S. Treasury, banks were so reluctant to start lending again.

The global financial markets may share various dynamics with robust-yet-fragile ecosystems, but what does it tell us, if anything, about how to avert or lessen the next crisis? After all, doesn't the RYF nature of the market mean that the risk of future calamity is built into the equation?

Increasingly that question is being asked not only by ecofinance theorists like Sugihara and Levin, but by leaders with more formal influence over global financial policy, including an unexpected voice in the world of risk management, working in one of the most traditional financial institutions in the world: Andrew Haldane, the executive director of financial stability for the Bank of England.

Like Levin and Sugihara, Haldane attributes the problems with the precrash financial order to the complexity of the system—the hornet's nest of interconnections between institutions—and the homogeneity of those institutions' business strategies. And his recipe for improving the resilience of the financial network bears striking resemblance to ecologists' prescriptions for ecosystems. "We need more complete, holistic measures of the health of the financial system and the dependencies between various institutions within it; we need to improve communications about it with the public at large in times of impending crisis or system flip, and we need to take steps to improve the financial system's biodiversity."

SEEING THE FINANCIAL WHOLE

Creating more holistic, EBFM-like measures of health for the global financial network begins with collecting a great deal more data. "Today, risk measurement in financial systems is atomistic—each institution, each node, is judged by itself," says Haldane. "After a systemic failure, this leaves policymakers—and indeed everyone—navigating in a dense fog."

To clear the air, regulators and governments need to be able to see the full number, size, and type of connections, flows, and dependencies between various institutions and markets at a glance. They need financial observatories to continuously replicate the Fedwire study, but on a much larger, more comprehensive, more timely, and more international scale.

Doing so will require building distributed sensor networks for the global financial system and bringing many more transactions into the light of day. This is already starting to happen. In the United States, for example, the package of financial reforms signed into law after the 2008 crash requires that derivatives like credit default swaps be cleared through mechanisms designed to mitigate the problem of counterparty risk. They call these mechanisms centralized counterparty clearinghouses, or CCPs.

In the precrash era, each bank dealt directly and independently with every other, so when the crash came one bank could never be sure if its counterparties had signed more contracts than they could afford to pay.

The clearinghouse addresses this murkiness by standing in the middle of every trade, becoming the buyer to each seller and then turning around and becoming the seller to each buyer. If one of the parties goes bankrupt, the CCP covers its side of the contract.

In theory, this achieves several things at once. First, it makes mapping the complicated web of relationships between financial parties much easier, as the clearinghouse sees every relationship between every buyer and every seller in its domain. "Clearinghouses compress the highly dimensional web of financial obligations to a sequence of bilateral relationships with the central counterparty—a simple hub and spoke network. The lengthy chain of relationships is condensed to a single link," says Haldane. Thus, in a crisis, a bank would not have to worry whom its business partners might also have dealings with—it would have dealings only with the clearinghouse. "Provided that link is secure—the hub's resilience is beyond question—counterparty uncertainty is effectively eliminated," he adds.

CCPs also simplify and cut down the thicket of counterbalancing claims between parties; if the First National Bank owes the Second National Bank $50 million, spread across a dozen derivatives contracts, and Second National owes First National a similar amount over a different set of contracts, the CCP can offset the claims through a process called netting. (In the wake of the financial crash, efforts were undertaken to do this manually in the CDS market by "tearing up" redundant, offsetting claims, reducing the volume of contracts by more than 75 percent. A centralized clearing facility would make this process automatic.)

Centralized counterparty clearinghouses also provide a mechanism to ensure that each party can actually pay out its contracts, because the clearinghouse requires each party to set aside sufficient capital to fulfill the contract as if it had to be settled each day—commonly known as marked-to-market accounting. Additionally, clearinghouses make it far less likely that any particular firm will become dangerously overexposed—as AIG and Lehman did—without anyone being able to tell, since the clearinghouse is in a position to see such buildups as they occur.

In addition to using central clearinghouses, other steps toward greater transparency have also recently been mandated in the United States, including the use of an open exchange for such derivatives, akin to a stock exchange. The primary benefit of exchange-based trading is in setting a market price. In the precrash era of direct deals between banks, nobody knew what anyone else was paying for these contracts. By moving to a market-based system, everyone can see what the going rate is, and companies are far less likely to get gouged.

Not everyone is happy with such an arrangement. Making derivatives trades more transparent in turn makes the underlying investment strategies of some investment firms more visible, and many of those strategies have previously depended for their success on being kept private. And Wall Street traders dislike exchange-based trading for another, simpler reason: When nobody knew what anyone else was paying for derivative contracts, they could charge whatever they wanted. The pricing transparency that comes with selling such contracts via exchanges naturally erodes their profits.

Nor is interposing a central counterparty clearinghouse in derivatives trading a panacea—it does not entirely eliminate the fragility in the financial network, but rather relocates it to the risk-management strategies of the central clearinghouses themselves. If managed well, a CCP can effectively mitigate the risk for a whole market; if not, its very centralized position could give it the starring role the next crisis.

But a centralized counterparty clearinghouse does enable the collection of vastly more information and greater transparency about the activities of the market. This data can be combined with many other sources to inform sophisticated, system-wide measurement tools that, much like EBFM, reveal the connectivity of institutions, not just their size and behavior.

In public health, practitioners often use informal social network mapping to identify and then inoculate super spreaders, those most at risk for initiating or propagating a contagion. So too in finance. "In early 2007, it's doubtful whether many of the world's largest financial institutions were more than two or three degrees of separation from AIG," says Haldane. "And in 1998, it's unlikely that many of the world's largest banks were more than one or two degrees of separation from Long-Term Capital Management, originators of the last major crisis. Mapping the links in the financial network might have identified these financial black holes before they swallowed too many planets."

Yet troublingly, prior to the crash, there was virtually no correlation between the size of the most important financial institutions and how much they held in reserve to cover potential problems. In spite of—or more likely because of—their supposed importance, these critical mega institutions seemed to have held less in reserve as a percentage of assets than their smaller peers. "One explanation is that, because big banks thought that they were diversifying, they thought they could afford greater risk," says Haldane. "Another explanation is markets allowed these banks to operate less conservatively because of the implicit promise of government support if things turned for the worse. Either way, there was no targeted vaccination of the super spreaders of financial contagion."

The best way to determine the appropriate levels of inoculation for such super spreaders is to model crises before they occur. This was undertaken in limited form in 2009 when the U.S. Federal Reserve began stress-testing banks that received bailout assistance, to make sure they had sufficient capital to weather difficult conditions such as sustained high levels of unemployment or ongoing home mortgage defaults. Stress testing gave a snapshot of each bank's health and risk exposure—and in the first round of testing, regulators discovered that more than half of the banks tested required additional capital. But even this limited stress testing is only a once-a-year affair, like an audit, and doesn't analyze how the system as a whole might be impacted if any of its subjects were to collapse. Such tests and simulations are regular occurrences for other network systems like electrical utilities, the military, and the air transportation system—and will need to become a regular feature of the financial system. And not just in a few market hubs, but everywhere.

Still, these are compensatory measures. They contain and minimize the damage of a systemic fragility once it has surfaced, but they can't flip the system back into its preferred state. Is there anything that can help the financial markets better self-regulate? The coral reefs may— once again—hold an intriguing clue.

THE BATFISH AND THE WIR

David Bellwood is obsessed with fish. His friends constantly poke fun at his ability to bring any conversation back around to them. When he was younger, growing up in northern England, he started an aquarium to observe reef fish. The interest propelled him all the way through an undergraduate degree at the University of Bath, where he did an honors project on the aquarium industry and then, later, to a PhD and professorship at James Cook University in Australia, focusing for more than twenty years on what is now his formal area of expertise and in-formal object of obsession: the parrotfish.

It's easy to see why Bellwood would fixate on this beautiful colored fish. The parrotfish is one of the mainstays of the coral reef system, and it can do some genuinely interesting things: Not only can it change gender—female parrotfish can transform themselves into males when their dominant male leader dies—but certain species of parrotfish have developed the ability to envelop themselves in a transparent cocoon, made from a viscous substance that comes out of an organ in their heads. This homemade sleeping bag disguises the scent of the parrotfish at night, leaving it safely hidden from nocturnal predators.

Of far greater interest to Bellwood, however, is the parrotfish's function on the reef: to consume large quantities of algae extracted from gnawed chunks of coral. Parrotfish are voracious algae eaters— Bellwood refers to them as "lawn mowers"—and on a healthy reef they play a critical role in regulating the competitive relationship between algae and corals. By cleaning the system of excess algae and enabling new corals to compete for space, parrotfish keep a coral-dominated reef in a continually self-regenerating state and prevent it from flipping into an algae-dominated state. Particularly after the difficult lessons of Jamaica, protecting the parrotfish and other herbivores has become a centerpiece of many reef resilience strategies.

Yet in recent research conducted on the Great Barrier Reef, Bellwood has discovered that the mechanisms that regulate the reef are more complex than marine ecologists had previously thought. His discoveries have significant implications for future reef management, and for other fields as well.

In controlled experiments, Bellwood and other researchers set up large open cages on top of the reefs—each one the size of a small office. In each of the cages they planted dense macroalgae assays, which simulated a reef system fully dominated by algae—what might be described as the parrotfish equivalent of an all-you-can-eat buffet.

Bellwood and his team set up multiple underwater cameras to film what came next. "We thought there was going to be a spectacular circus because we had forty-five species of herbivorous fish in the vicinity, and huge densities of parrotfish. This particular location should have brought a feeding frenzy to the algae."

The team sat waiting in anticipation for the first few hours. That anticipation soon turned to bafflement. The only thing they could see on the video footage was a thick cluster of algae waving in the current. Not only were there no parrotfish, there were no fish to be seen at all.

"Lights! Camera! Action! And then . . . nothing. Were the fish reproducing or gone somewhere else? I mean, this was three-meter-high, beautiful algae. These are herbivores. These fish should have been having a feast."

What the team did notice, over the next few weeks, is that the algae were slowly getting *thinner*. "You know how people get older and you can slowly start to see their scalp? Well, the algae were like that. You still had the bits that were reaching to the top but it was getting really thin in the middle. Eventually, there was very, very little on the footage."

Bellwood was expecting the algae to be gobbled down in a frenzy, primarily by parrotfish, in twenty-four hours. Instead, over a period of three weeks, the algae were slowly nibbled and eroded until they finally just collapsed—but what was eating them? The results were so curious that Bellwood and the other scientists didn't even have a framework to understand them. All they could do was go back and review the footage again and again.

Then, during one of their repeated viewings, they spotted something altogether unexpected. From out of the murk, a completely unrelated species, a small black fish with a golden fringe (called a pinnate batfish) appeared on the footage. And then, to everyone's shock, it slowly started feeding on the algae.

"We were stunned for two reasons. To begin with, batfish aren't supposed to eat algae. They are an invertebrate feeder, not an herbivore. Secondly, we could never catch them doing it. As soon as we got in the water, they would swim away. It was like the *Far Side* cartoon where the cows are having a conversation and then the car comes and they all suddenly go back to eating the grass."

Bellwood's research suggests that while the parrotfish act as lawn-mowers for the reef, they can do so only when the reef is in the healthy, coral-dominated state. When the system has flipped and algae have taken over, they're no longer able to provide this function. And that's when the batfish—which normally doesn't eat algae—is "deployed" on the reef to correct the imbalance. One prevents a flip, the other reverses it.

Bellwood likens the reef to a golf course. "Under ordinary circumstances, the equipment you need is a lawn mower because it keeps the nice grass down. The coral reef lawn mowers are the parrotfish, surgeonfish, and other species that graze the turf and keep the greens tidy. But if, for some reason, your mowers are broken, the weeds on the golf course start to get big, and soon there is a back garden overgrowth.

Now, by the time the lawn mowers show up, they won't work. You need a chainsaw and a bush saw. Surprisingly enough, we discovered that this is how the batfish functions."

The batfish are part of what Bellwood calls a "sleeping functional group," a species or group of species capable of performing a particular functional role—but which do so only under exceptional circumstances.

Bellwood's team was both euphoric and confused by the discovery of these sleeper groups. The good news is that fish exist with the capability to help reverse a system flip from an algae-dominated state back to a coral-dominated state. The bad news, however, is that in many cases, scientists have no idea which fish they are. Because the resilience function of these species emerges only in exceptional circumstances, no one before Bellwood had ever connected the batfish with coral-algal interactions in more than fifty years of research.

What might it look like for the financial system to have such sleeper functional groups of its own—countercyclical strategies lying dormant within the financial network, stirring only when the system flips, as it did in 2008? Finance may just have found its batfish in tiny Switzerland, in the form of a unique alternative currency called the WIR.

If you have traveled around Switzerland at any point in the last few decades, you may have noticed stickers in the windows of local shops and businesses: "We Accept WIR." Perhaps you caught sight of one of WIR's glossy catalogs displaying a full list of all of its participating businesses. If you stayed in a hotel, the clerk may have even asked you if you wanted to pay through WIR. No doubt she or he soon rescinded the offer upon realizing that you were not a native of Switzerland.

What is this mysterious alternative currency and how does it work?

Bernard Lietaer, a Belgian economist with more than twenty-five years of expertise in currency systems, has been a keen observer of this Swiss alternative currency for many years now. He is not alone. A number of macroeconomic analysts are starting to look more carefully at the way it functions.

WIR—*Wirtschafstring,* or "circle" in German—began in the depths
of the Great Depression. In the wake of the stock market crash of 1929,
total world trade plummeted—by 20 percent in 1930, another 29 percent in 1931, and another 32 percent in 1932. Unemployment reached
30 million. Wealth that was once all but assured disappeared into thin
air, and banks that seemed certain to stand suddenly collapsed, as stocks
lost nearly 90 percent of their value.

Switzerland was pulled into the crisis more slowly than some of the
other European countries, and it was slower to recover. By 1934, when
the United States and Germany were showing faint signs of recovery,
Switzerland remained in a malaise. The number of bankruptcies hit
record heights and trade and tourism took deep hits. The Swiss railroad,
SBB, carried a deficit double that of the federal budget. Unemployment
was the norm, not the exception: In 1934, there was one job opening
for every seventy-three official job seekers.

In the midst of all this, sixteen businessmen decided to create a
solution for themselves. All of them, along with their clients, had been
informed by their banks that their credit lines were no longer open to
them; without credit, their businesses were positioned on the brink of
bankruptcy. Rather than fail, they decided to set up a complementary
form of currency.

Their result, the WIR, is a mutual credit system. A debt in WIR is
either reimbursed by bartering in sales with someone else in the network or paid in full in the national currency. Over time, this network
expanded to include one-quarter of all the businesses in the country.
Today it is a thriving barter network that makes up a well-recognized
complementary currency in Switzerland.

An analysis of more than sixty years' worth of data on the WIR by
macroeconomist James Stodder has shown that whenever there has
been a recession, the volume of business in this unofficial currency has
expanded significantly, cushioning the negative impact of lost sales and
increased unemployment. Whenever there has been a boom, business
in national currency has boomed, and activity in the unofficial currency

has dropped proportionally again. In the past, people have attributed the success of the Swiss economy to a national character of pragmatism and thrift. Stodder's study offered unexpected proof that the secret behind the country's legendary stability and economic resilience is the spontaneous countercyclical behavior of this small alternative currency system.

WIR functions precisely like Bellwood's batfish—a latent, just-in-case contingency system that is activated when the economy is at or near a phase shift. Lietaer is an advocate for more complementary currencies that function like the WIR—business to business—in the European Union and the United States. "The substance that circulates in our global economic network—money—is maintained as a monopoly of a single type of currency (bank-debt money, created with interest). Imagine a planetary ecosystem where only one single type of plant or animal is tolerated and officially maintained and where any manifestation of diversity is eradicated as an inappropriate 'competitor' because it is believed it would reduce the efficiency of the whole."

The batfish and the WIR illustrate, in very different ways, an essential strategy of many resilient systems: They are embedded countercyclical structures that can respond proportionally, and in the same time signature, to disruptions as they emerge. These structures are part of an essential inventory of diverse tools often found in resilient systems: The batfish is an example of biodiversity, the WIR, of economic diversity. Like all inventory, this diversity imposes a carrying cost. And because such structures are latent in a system—they are only called upon when a crisis emerges—it can be difficult to place a value on them when things are in a more humdrum state of affairs. In a relentless quest for greater systemic efficiency, this diversity can be lost and unmourned—until it's too late.

As we've seen, the fragility and resilience of most systems begins with their structure. The complexity, concentration, and homogeneity of a system can amplify its fragility; the right kinds of simplicity, localism, and diversity can amplify its resilience. The lens of resilience suggests, for example, that what's needed is a smaller, simpler, more accountable, and more decouplable financial system, with genuinely diverse participants, that is more closely aligned with its original purpose—providing liquidity to organizations and individuals—than with its more recent one, which seemed to be engineering wealth from thin air.

But resilience takes more than just the right structure—it takes the right kinds of processes and practices: measuring the whole health of the system, like EBFM; modeling and stress-testing the system, like the Fedwire study; scanning for emerging disruptions and mobilizing the right, inclusive responses when they strike, like the proposed financial observatories; building in feedback and compensatory systems, like the batfish and the WIR, that help keep the system in check. Most of all, it requires keeping the system dynamic and reconfigurable. As we'll see next, it is in such dynamism that many systems' resilience rests.

2

SENSE, SCALE, SWARM

Resilience isn't just found in systems we admire, but sometimes in systems we loathe. Terror networks and even many diseases can be thought of as perversely resilient—they persist and even thrive under sustained and powerful assaults intended to root them out and eliminate them. How do they do it? And how might their strategies for doing so be applied in more positive contexts?

As we'll see, the answers are often rooted in the dynamics of these kinds of entities. For example, terror networks and many microbial infections survive, in part, by modulating their "metabolism" up and down—reducing their operations to near dormancy for long periods, then, when they sense the time is right, scaling up to attack in highly coordinated swarms. Applied in very different circumstances, these sensing, scaling, and swarming tactics hold promise for improving the resilience of systems on which civilization more positively depends.

OPERATION HEMORRHAGE

In November 2010, an unusual document began to circulate in a few less savory corners of the Internet—a digital, full-color magazine that, at a casual glance, might easily have been mistaken for an industry trade publication. And, in a manner of speaking, it was: This was the third issue of *Inspire*, the English-language magazine produced by al-Qaeda in the Arabian Peninsula (AQAP), the branch of the global terror network largely based in Yemen.

Produced by a radicalized American convert to al-Qaeda named Samir Khan, issues of *Inspire* are filled with a mix of propaganda aimed at young American and British citizens: advertisements for jihad, home-brew recipes for terrorist activities (sample headline: "Make a Bomb in the Kitchen of Your Mom"), and fawning profiles of al-Qaeda leadership. The magazine's writing was naïve, youthful, and earnest, with countercultural flourishes intended to speak to disaffected Western teenagers in a language they understand.

Issue 3 of *Inspire* had a slightly different focus, however, as signaled by its cover, a blurred photo of a UPS cargo plane emblazoned with a bold, singular headline: "$4200."

The dollar figure was a reference to the total cost of an AQAP terrorist plot, entitled Operation Hemorrhage, which was intended to blow up two cargo planes—one operated by UPS, the other by FedEx—carrying packages bound for the United States. As described in exhaustive detail in the magazine's ensuing pages, AQAP bomb makers had carefully hidden high explosives inside two hollowed-out desktop printer cartridges; the cartridges had been connected to detonators and circuit boards lifted from mobile phones, then repackaged with their original printers in a way that made them appear factory sealed. These were then shipped by AQAP operatives from Sanaa, Yemen's capital, addressed to liberal Jewish synagogues in Chicago, President Obama's hometown.

The fictional names of the bomb's recipients, Reynald Krak and Diego Diaz, were derivations of historical figures from the Crusades and the Spanish Inquisition. In a further literary grace note, the bombers also enclosed a copy of Charles Dickens's *Great Expectations*—a reference both to their optimism about the plot's chances of success and a coded message of sorts: Anwar al-Awlaki, AQAP's chief theoretician and religious leader, had become a Dickens fan while in prison in Yemen, when he had been banned from reading Islamic texts. (Al-Awlaki has since been killed in a drone strike, as has Kahn.)

These rhetorical flourishes were hardly the main point of the plot, however. As the issue of *Inspire* goes to lengths to explain, Operation Hemorrhage had two central aims: one strategic and one economic. First, the bombs would act as a sensor: They would pass through the latest air cargo security equipment, providing AQAP with good intelligence regarding the West's explosives detection capabilities. Second, the bombs would be a provocation: The ensuing security fears would force the West to invest billions of dollars in new security procedures:

> From the start our objective was economic. . . . The air freight is a multi-billion dollar industry. . . . For the trade between North America and Europe air cargo is indispensable and to be able to force the West to install stringent security measures sufficient enough to stop our explosive devices would add a heavy economic burden to an already faltering economy.

A few months before, AQAP had prototyped its attack, deploying a bomb of similar design to bring down a UPS cargo plane leaving Dubai International Airport, killing the two pilots aboard. Counterterrorism officials had not registered AQAP's involvement at the time, instead attributing the crash to mechanical failure.

This time, however, Saudi Arabia's intelligence agency caught wind of the attack just as it was coming together and alerted officials in the United Kingdom and Dubai that the packages were being shipped by air from Yemen to the United States by way of Europe. Security officials seized one of the two bombs en route at the East Midlands

Airport, in Nottingham, UK, on a UPS plane stopping over on a flight from Cologne to Chicago; the other was found at a FedEx facility in Dubai. Although not independently confirmed, one of the two bombs was defused just seventeen minutes before it was set to explode, according to France's interior minister.

Yet despite the operation's apparent failure, AQAP had deemed the initiative a great success:

> Two Nokia mobiles, $150 each, two HP printers, $300 each, plus shipping, transportation and other miscellaneous expenses add up to a total bill of $4,200. That is all what Operation Hemorrhage cost us. In terms of time it took us three months to plan and execute the operation from beginning to end. On the other hand this supposedly "foiled plot," as some of our enemies would like to call, will without a doubt cost America and other Western countries billions of dollars in new security measures. That is what we call leverage.

Leverage: In the logic of post-9/11 terrorism, it's priceless. AQAP didn't even have to blow up a plane to achieve its aims—all it needed to do was provoke a costly response.

Indeed, Operation Hemorrhage's value as a media activity may have even outweighed its value as a terrorism activity. In open sourcing and publicizing its methods, AQAP advertised both its own imperviousness and the relative vulnerability of its chosen enemy, all the while encouraging others to emulate its approach. At the same time, it also reinforced its own metanarrative about the conflict and its role within it—engaging in a battle to control the story. That's where those rhetorical flourishes, like the Dickens novel, came in: They're the sorts of sticky, surprising details that stand out and pass virally from person to person, further drawing in potential recruits. Evidence of these tactics' potency already exists: When nine young men in the United Kingdom were arrested in December 2010 and charged with plotting to blow up the London Stock Exchange, Big Ben, and the U.S. embassy, authorities found them carrying two issues of *Inspire* in their pockets.

In the decade since 9/11, terror networks like AQAP have been on the receiving end of the largest sustained military eradication efforts in modern history. A conventional military force a thousand times its size would have long since been defeated. During the same period, however, even with the recent death of Osama bin Laden, al-Awlaki, and other figureheads, al-Qaeda-affiliated or -inspired chapters have spread to more than sixty countries around the world. Rooted in its networked structure, its ability to modulate its operational metabolism, and its ability to swarm, al-Qaeda's success holds powerful lessons for designing positive forms of systemic resilience in domains far beyond terrorism.

For several decades, Dr. John Arquilla, professor of defense analysis at the U.S. Naval Postgraduate School, has been one of the world's leading thinkers on the resilience tactics of groups like al-Qaeda and the real-world dynamics of netwar, a term he coined in the 1990s with colleague David Ronfeldt to describe what they predicted would become the twenty-first century's dominant mode of conflict.

"Organizational forms profoundly affect a variety of endeavors, whether peaceful or warlike," says Arquilla. "Assembly line–driven organizations, for example, manufacture using certain processes, and their management structures come to echo those processes. The 1950s-era corporate management hierarchy and the military's current command-and-control model are both mirrors of their institutions' basic operations.

"About two decades ago, it became clear that networks were going to enable new organizational structures on the battlefield, and indeed a whole new way of warfare. This would have enormous implications for both the structure and tempo of military operations, because the way you and your enemy organize determines the kind of war you fight."

The first important element of al-Qaeda's success has been the rise of its network franchise model, in which al-Qaeda has become less and less a formal organization and more a global organizing principle and open-source brand. This brand has proved alluring to local groups with very different but overlapping objectives. In exchange for affiliation, such groups became part of a larger movement that amplifies local perceptions of their global power and prestige. Thus, the Salafist Group for Preaching and Combat, based in Algeria, which had previously been focused on overthrowing that government, became al-Qaeda in the Islamic Maghreb; a violent Islamist splinter cell in Indonesia became Al Qaeda in the Malay Archipelago. In addition to providing the larger network with global reach, these various groups also specialized, based on their history, makeup, and unique skill sets. AQAP, for example, emerged as a leading recruiter of Americans to join the global jihad movement—*Inspire* being just one manifestation of their efforts.

This loosely affiliating, self-organizing, networked, and self-specializing dynamic is replicated inside terrorist networks and other asymmetric fighting forces as well as between them. In such arrangements, small groups are not bound together via traditionally strong command-and-control structures, but by ad-hoc, redundant, and informal social connections—less like the Marines and more like a pickup game of basketball. The small scale of the groups within such networks helps them remain agile, while the many-to-many ties in the larger network ensure that even if 10 percent or 20 percent of its membership is eliminated, the network as a whole will continue to function. "How many times have we killed number three in al-Qaeda? In a network, *everyone* is number three," notes Arquilla, dryly.

Terror networks succeed by redefining battlefield success downward—their goal is obviously not to win in traditional military terms, but to embarrass and exhaust their enemies. Battling a traditionally far superior force to a draw counts as victory. To achieve this, terror groups balance long periods of near dormancy with intermittent high-energy bursts of activity, like Operation Hemorrhage. The purpose of such operations is not merely to wound the enemy, but to incite an ineffective response that will make him appear all the more impotent.

This was a central dynamic in the 2006 Lebanon War, when somewhere between one and two thousand Hezbollah fighters held their own against more than a hundred thousand Israeli troops and a relentless air campaign. Notes Arquilla: "On the first day of the war, Hezbollah launched two hundred rockets; on the last day of the war they launched two hundred rockets. Most Israelis and most people around the world feel that Hezbollah won the war, just by hanging in to the end. And how did they win it? By dividing up into several hundred little firing teams of three to four individuals, marching about, firing off hidden Katyushas, and melting back into the scenery. We call this strategy 'shoot and scoot.' And it's the defining structure of the conflicts in Afghanistan, Iraq—increasingly, everywhere."

When not fighting, very little can distinguish unconventional fighters or terrorists from local civilians— particularly where they share strong local, cultural, religious, and even family ties—making them incredibly challenging to detect and eliminate. This ability to scale down rapidly has other benefits as well: It reduces the group's support requirements, it preserves their future fighting capacity, and it increases the likelihood that their more traditionally organized opposing force might accidentally kill or otherwise offend innocent civilians— mistakes that can be disastrous to the larger effort to frame the narrative of the conflict.

Yet, though dangerous, terrorist cells must at least occasionally "delurk" to complete missions like Operation Hemorrhage. Doing so is extremely dangerous, to be sure, but such operations serve as necessary advertisements for the group's potency and validate its relevance to potential recruits. More important, they serve to keep the enemy off balance, which is critical to the terrorists' preferred method of conflict. "By keeping its chosen enemies sporadically and schizophrenically engaged, the global Al Qaeda network hopes we'll 'punch ourselves out,' much like Muhammad Ali did to George Foreman," says Arquilla.

In the United States, politicians often argue about why al-Qaeda has not staged another 9/11-style spectacular attack on U.S. soil. It may be that, having induced the United States into a decade-long, trillion-dollar conflict with it and its allies, on its terms, it simply doesn't need to. Our continued, imperfect engagement suits its long-term, rope-a-dope purposes perfectly.

To improve the effectiveness of the attacks they do undertake, terror networks have increasingly adopted a new offensive technique that Arquilla calls "swarming." In this model, small highly distributed teams simultaneously attack nonmilitary targets, overwhelming defenses—if there are any—originally designed to deter a single large aggressor. "In the 2008 Mumbai bombing, just ten attackers in five two-man teams struck at the same time—five different places in one city. They completely overwhelmed a very wealthy and pretty well militarized country—India—and held a major world city hostage for two days, killed two hundred people, and caused untold disruption upon that society. That's the template of what's to come."

LEARNING FROM TB

In their ability to rapidly scale up and down, terror networks bear surprising similarities to another highly resilient but altogether different phenomenon: the pathology of tuberculosis infection in the human body. These similarities suggest both lessons for designing resilient systems and ways of contending with perversely resilient phenomena.

TB is one of the most prevalent diseases on Earth. One in three people carry TB antibodies, meaning they have been exposed at some time in their lifetime, but only a small fraction—about 10 percent of those exposed—will ever become actively sick with the illness. Still, TB's toll is staggering: Every day, about 4,700 people, mostly poor Africans and Asians, die from the disease. After HIV, it remains the most common infectious cause of death in adults worldwide. New infections occur at the rate of about one per second—two new ones occurred while you were reading this sentence.

TB bacteria enter the body through the air, carried in invisible, microscopic droplets to the alveoli, the tiny air sacs of the lungs, where they take up residence. Within approximately six weeks, a newly infected person will develop a small infection, called a primary infection, rarely accompanied by any symptoms.

At a cellular level, however, something truly remarkable is happening. When the TB bacteria arrive in the lung, they are met by *macrophages* (literally "big eaters"), the white blood cells at the front lines of the human immune system that are normally responsible for consuming and destroying invading pathogens. In certain cases, macrophages find this task difficult to do, so in a last, kamikaze-like move, they will engulf an unknown invader, coat it in the cellular equivalent of Saran Wrap, and then, on cue, promptly die, taking the pathogen with them. However, in TB, occasionally just the opposite occurs: TB takes over some macrophages, preventing them from dying, and turns them into zombie-like incubators for producing more TB bacteria, which slowly replicate inside until they burst their cellular hosts open and spread to others.

This process isn't perfect—many TB bacteria are eliminated by untainted macrophages, which sense the distorted chemical signature of their TB-infected counterparts and can turn on them. In more than 90 percent of infected people, the hosts' immune response is just effective enough to control, but not entirely eliminate, the TB infection, and the disease enters a long period of latency, which can last for years.

"This long latency period is part of the reason TB is so insidious," says microbiologist Sarah Fortune, a TB researcher at Harvard. "In its latent state, TB doesn't do very much, which is why our traditional tools against it, antibiotics, are not terribly effective. After all, antibiotics target metabolism, and with latent TB, there isn't much being metabolized."

Paradoxically, TB's long latency is also one of the things that make it so deadly. It allows TB's human hosts to grow up and reproduce, creating a new generation of people for the disease to spread to. "TB is like a symbiont that occasionally kills you," says Fortune. "In contrast, illnesses like Ebola are far rarer, precisely because they become deadly so quickly. In a matter of weeks, all the human hosts are gone, and they take the outbreak with them."

While the precise biomolecular mechanics are still not perfectly understood, the current model of TB's pathology suggests that during the latent phase of the disease, TB bacteria constantly probe the immune system for weaknesses, much like the AQAP terror cell probing the global security system. "Our reasoned assumption is that during the latent phase, some TB is active, much is inert, and then, some critical threshold is passed, often in an immune system compromised by other illness such as HIV, alcoholism, or diabetes, and the illness enters its active phase," says Fortune.

At that point, everything changes. In the active phase, the TB infection is aggressively reactivated throughout the body, though the lungs are the preferred location for the illness to strike. Once there, in an eerie parallel to terror networks, the aggressive TB bacteria depend on provoking an overreaction from the immune system in order to spread.

This overreaction is expressed as a caseous granuloma, which forms when a ball of TB bacteria and TB-infected macrophages collect in the lungs, where they are attacked by healthy immune cells. Like cops surrounding a building with a gangster inside, these cells employ a standard immunological strategy: They try to wall off the TB within a fibrous shell.

Tiny granulomas are a normal part of the immune response and form throughout the latent phase of the disease. When they form during the active, aggressive phase, however, things can quickly go haywire. This time, when immune cells surround the ball of TB cells, they die, in turn attracting even more immune cells, which attempt to wall them in. Then *they* die, in a process that repeats over and over again, inflating the granuloma like a balloon.

Inside this expanding shell, the TB bacteria sit amid clusters of now-dead immune cell tissues. Eventually, such granulomas (called "caseous" because of the cottage cheese–like consistency of their dead centers) can grow to be the size of a tennis ball inside the lung. And as they grow, the necrotic material at their core begins to liquefy, forming the perfect environment for TB bacteria to multiply. And the TB does just that: A modest 2-centimeter granuloma alone can contain 100 million active bacteria.

As the granuloma grows, it irritates the lining of the lungs, eventually bursting and spraying the inside of the lungs with the bacteriological equivalent of a swarm attack, hitting many targets at once. "TB depends for its survival on inducing this tactical error on the part of the immune system," says Fortune. "The effort to contain the infection serves only to concentrate and then amplify it."

After that, every time the actively infected person coughs, he or she can release untold millions of invisible TB bacteria into the air. An average infected person can cough up to thirteen infectious doses an hour; a severely contagious one, up to sixty—before, in about half of all untreated active cases, they finally pass away.

EMBRACING THE SWARM

Both TB and terror networks maintain their resilience amid sustained attack by moderating their metabolism, scaling down to near dormancy for long periods of time, and scaling up to strike when the time is right. Both depend on mechanisms for probing their target's responsiveness and dynamically reorganizing at the right moment. Both spread by provoking an overreaction from their host targets. And both succeed by swarming in coordinated, simultaneous attacks. These similarities are starting to suggest new, biologically inspired metaphors for tackling terror networks.

For Arquilla, the answer to tackling the threat of terror groups is to absorb and mimic their tactics: to meet the network with a network.

"The War on Terror is actually the first great war in which we're seeing nations at war with networks. The United States and its allies initially approached things in very traditional ways: massive deployments, overwhelming force, and shock and awe." Like an incomplete course of broad-spectrum antibiotics, these approaches were sufficient for clearing the battlefield but not for holding it. And as they failed to do so, they have slowly, painfully, been replaced with lighter, more agile, targeted and networked approaches.

In an American corollary to Hezbollah's shoot and scoot model, Arquilla describes what finally came to work in the U.S. military's campaign against networks in Iraq, a strategy he calls "outpost and outreach": "Several years into the war, over much resistance, we moved our soldiers from huge bases to hundreds of small outposts thirty to fifty soldiers apiece, dramatically expanding the number of nodes in our physical network." This quickened response times, allowing U.S. troops to respond speedily, flexibly, and proportionally to a threat or problem, within minutes instead of hours. It also encouraged real relationships with local residents, creating connections that generated huge amounts of actionable intelligence.

Complementing this physical reorganization was an outreach and social networking effort that is widely regarded as yielding the greatest tactical success of the war: nurturing the Sunni Awakening movement, a coalition of Sunni tribes that were induced to stop fighting the Americans and start fighting al-Qaeda's local branch instead. "The combination of the physical network and the social network—that's really what turned the campaign around on the ground. And we couldn't have gotten that to happen without effectively engaging in a battle for the story, with our own counternarrative."

Supported by the success of outpost and outreach, U.S. military forces have also started to embrace swarm tactics of their own, aimed at excising local terror groups whole. Like the slow, persistent efforts used to take down narcotics and Mafia networks, these counterterrorism and counterinsurgency operations are predicated on long wait-and-watch intelligence-gathering efforts, which gradually illuminate latent ties in the terrorist's social network. Then, when as much of the network has been revealed as possible, agile military teams take it down in coordinated, swarming strikes.

Like the use of targeted antibiotics to tackle TB, the timing, speed, simultaneity, and precision of these strikes is important. Because many-to-many networks are naturally resilient, normal military kill or capture strategies are too slow to ever be effective—the loss of individual members is simply routed around by those remaining, like packets routed around downed computers on the Internet.

"This presents a huge and as yet unfinished cultural change for military forces, which have an inherent doctrinal preference for rapid destruction," says Arquilla. "We're slowly realizing that to win, you have to preserve as much information as possible, for as long as possible, before striking. And that means not blowing things up, at least until the last possible minute." In netwar, as in immunology, the timing of force is often more important than its scope.

These efforts, though effective, are still a retrofit of existing military practices. What would it look like for U.S. forces to embrace fully the self-organizing dynamics of its networked adversaries? Arquilla suggests the military may one day adopt the market-driven dynamics of companies like eBay: Instead of issuing pages and pages of operational orders, a commanding general could post tactical objectives to a website and assign points values to them: 100 points for this bridge; 500 for the capture of this town or that enemy combatant.

"The general's various units would log on, see what's listed, and bid for whatever is there. If there was something the general wanted done that wasn't getting done, he could raise the point values. Or lower them. And, at the end of the day, if there was something that still wasn't done, he might have to give an actual order to someone. He could still do that. But I believe we'd see self-synchronizing campaigns that are faster, more adaptive, and more effective."

And what would the fighting force look like that wages such

campaigns? Arquilla envisions an army of swarmers. "I would rest easy if twenty years from now, instead of ten army divisions, we had a hundred clustered small units," he says. Like its enemies, this military would structure itself to scale up and down more quickly. "We have more than two million people right now; I think we could take it down to under four hundred thousand in the active force and have a larger reserve, but just that: a reserve. We could achieve all of this at about half the cost that we're paying today, if we take networks seriously. It's not rocket science. It's network science."

Just as the U.S. military is finding increased resilience and tactical success in embracing some features of its more agile networked adversaries, other big systems are being similarly transformed by these highly distributed, scalable, and networked approaches. Nowhere is this being felt more than in the slow reengineering of the United States' massive electrical grid, and in the planning for the future system that will one day replace it.

THE GRID THAT BREATHES

Referred to by the National Academy of Engineering as "the supreme engineering achievement of the 20th century," the modern grid powering North America is larger and more ubiquitous than the domestic Internet, and just as complex. In the modern grid, electricity generated from all forms pours in from power plants onto an interconnected highway of weblike regional grids that serve areas hundreds or thousands of miles across. This system of systems winds its way from one end of the United States to the other, while crisscrossing back and forth through Canada and Mexico.

Because electrical current follows a path of least resistance and cannot be stored, its journey is not like a car idling at a stoplight, but more like water pouring through a network of gated sluices at light speed. (Most electricity is consumed less than a second after it has been generated.) If there is a traffic jam on one transmission line, electricity is rerouted to its final destination in ways that can be surprisingly circuitous. When the lights are turned on for dinner in Portland, Oregon, for example, some of the electricity often whizzes from Los Angeles and then takes a detour through Utah before lighting the bulbs up north. When the TV is powered up for *Sunday Night Football* on the East Coast, the electricity might travel, at the speed of light, from Canada and down through Ohio and even Virginia before linking back up to deliver the game in upstate New York.

The system moving all these electrons around is not just technological. It is characterized by tight couplings of human organizations and businesses, a system so big, more than 20 percent of all electrical infrastructure purchases on Earth are used just to keep the North American grid operating. Take a moment to consider the system's vast breadth: According to a 2005 analysis by the Global Environment Fund and the Center for Global Smart Energy, more than 3,200 electric utilities and 2,000 independent power producers serve 120 million residential household customers, 16 million commercial customers, and 700,000 industrial customers. They use the continent's 700,000 miles of high-voltage transmission lines, owned by about two hundred different organizations and valued at more than $160 billion. In turn, more than 5 million miles of medium-voltage distribution lines and 22,000 substations, owned by more than 3,200 organizations and valued at $140 billion, plug into these high-voltage superhighways.

Now catch your breath. Did you take all of that in? Probably not. The system is too complex for most people to comprehend in its entirety. And just describing the connections between the nodes on this vast network captures only one aspect of its performance. For example, the system breathes with the seasons: The electrical conductivity of the grid's physical components—and therefore how much energy needs to be transmitted—changes in hot and cold weather, as their materials sag or contract, like a diaphragm.

The thousands of human beings who make sure the lights stay on have to constantly broker the mission-critical need of humanity for electricity, an electrical production and distribution system of mind-boggling complexity, and the weather. Doing so is an art akin to musical performance, requiring an understanding of the time of day, the day of the year, the geography of the region, major social events (like a rock concert at a local arena), and countless other variables that have nothing to do with the equipment itself.

To make matters worse, a low-probability, high-impact wild-card event like a heat wave or a cold snap can send electrical demand soaring instantly, overwhelming power lines. Similarly, mechanical failure like a downed transmission line or a lightning strike can cause a surge that, if not disposed of correctly by either automatic systems or human operators, can threaten vast regions of the grid itself.

And that's just what happened on August 14, 2003, shortly after 4:00 p.m. Hot summer temperatures in the northern and central regions of the United States prompted more and more people to crank up their air conditioners and fans. In the heat of the summer, several local power lines in northern Ohio started to expand and sag, brushing up against overgrown trees that needed trimming, creating a short circuit and shutting down the power in a cascade of sparks.

On an average day, it would have been a very low-level disturbance, setting off an alarm in a local utility's control room and alerting one of the local operators to the disturbance. Under normal circumstances, that operator would simply reroute the power flows to avoid the site of injury, and a crew of linemen would be dispatched to make the physical repair.

August 14, 2003, however, was not an average day. In addition to the shorted power lines, the alarm software that would have typically notified the local operator of the problem also failed. Unaware of the damage, all of the other operators wheeling power across the regional grid continued to route electricity through the damaged area, forcing the transmission lines to bear an untenable amount of current. Stressed well beyond their capacity, within two hours of the initial short circuit all of the power lines in Ohio cut out entirely.

Just like other robust-yet-fragile systems described earlier, the power outage in Ohio spread from a small triggering event to a systemwide collapse, tripping circuit breakers and sending more and more plants and lines offline. The grid operators were caught completely by surprise and, equipped with outdated and incompatible monitoring equipment, were unable even to see what was happening. Within eight minutes, the blackout affected an estimated 10 million people in Ontario and 45 million people in eight U.S. states. Bigger than 1965, much bigger than 1977, the blackout of 2003 was the largest power outage in North American history. As day turned to night, air and train travel on the East Coast ground to a halt. Cleveland declared a curfew on all persons under the age of eighteen. In some affected urban areas, the Milky Way became visible for the first time in decades.

All this, triggered by a tree branch and some by-now-familiar factors: systemic complexity, lack of transparency, and a lack of interoperability.

TOWARD A SWARMING GRID

The electrical grid has always had at best incomplete self-monitoring capabilities. In the late 1920s and early 1930s, early analog computers were built using electric components to reproduce miniature electric systems. Each of these early devices—big enough to fill a living room—would simulate the impact of disturbances to the grid. By the 1960s, however, the new digital computers allowed for a more sophisticated system. Most power plants and transmission lines switched over to something called SCADA, an acronym that stands for "supervisory control and data acquisition." These systems collect data from various sensors at a plant or in the lines and then send it off to a central computer that manages and controls it.

Like many protocols, the interconnectedness that came from using such a standard initially resulted in greater efficiency. But over time, as the physical grid and its associated technologies grew in sophistication,

SCADA systems proved incapable of capturing the fuller state of the grid and had the effect of slowly eroding, not supporting, its resilience.

According to Phillip Schewe, with the American Institute of Physics, this kind of technology is too old, too outdated for today's challenges. It "does not sense or control nearly enough of the components around the grid. And although it enables some coordination of transmission among utilities, that process is extremely sluggish, much of it still based on telephone calls between human operators at utility control centers, especially during emergencies."

Sometimes even these telephone calls between human operators prove useless because neighboring utilities can use incompatible control protocols. A collection of transcripts gathered on the day of the 2003 blackout illustrates the lack of interoperability and real-time information between the system managers, as in this example, between operators of the Pennsylvania–New Jersey–Maryland and Midwest Independent System Operator regional grids:

> PJM: It looks like they lost South Canton–Star 345 line. I was wondering if you can verify flows on the Sammis-Star line for me.
>
> MISO: Well, let's see what I've got. I know that First Energy lost their Juniper line, too.
>
> PJM: Did they?
>
> MISO: And they recently have got that under control here.
>
> PJM: And when did that trip? That might have . . .
>
> MISO: I don't know yet . . .
>
> PJM: And right now I am seeing AEP systems saying Sammis to Star is at 1378 . . .
>
> MISO: Let me see. I have got to try and find it here, if it is possible . . . I see South Canton–Star is open, but now we are getting data of 1199, and I am wondering if it just came after.
>
> PJM: Maybe it did.

The operators were reduced to managing the "greatest engineering triumph of the 20th century" using technology invented in the nineteenth—the landline telephone. Their situational awareness was incomplete, murky, and based on old information.

Simply patching these problems wouldn't make the grid any more secure: Every day, the grid gets larger, more complex, and is tasked with doing things its inventors could scarcely have imagined. And a darker concern lurks: While this blackout has been an accident, what if the next time the malfunction was intentional?

Working on behalf of the power companies, hackers had already discovered security vulnerabilities in SCADA systems: In one simulation, they remotely hacked into the controller typically used by grid operators, surreptitiously recording all of the outbound instructions sent to various generators and monitoring the responses. Then they simply switched the instructions normally sent during the day, when demand is at its peak, and those sent at night, when much of the system is idling, succeeding in crashing the (thankfully simulated) grid in a matter of minutes. A modified version of this approach has now been confirmed in the wild: On November 8, 2011, hackers are suspected of remotely gaining access to a water pump's controller connected to the Springfield, Illinois, municipal water supply. By sending commands to rapidly turn the pump on and off, they burned it out, destroying it in the process.

Even without the threat of cyberterrorism, the likelihood of repeats of the 2003 blackout have been increasing, driven by the ever-increasing demand for larger power transfers over longer distances, underinvestment in the transmission system, wild-card spikes in demand, and the financial consolidation of power utilities. Then there's the anticipated wave of coming failures among aging electrical transformers, installed in the 1950s, '60s, and '70s and now rapidly approaching the end of their useful lives. Throw into the mix the increasing frequency of climate-change-induced anomalous weather events and the fact that U.S. drivers will soon be plugging hybrid electric vehicles into the grid by the millions, and it becomes clear: If we are ever to ensure that the grid meets the growing needs of the twenty-first century, consistently, safely, and securely, the grid will need to be transformed—but how?

Enter engineers like Massoud Amin, widely considered the father of the smart grid. Amin, a stately Iranian immigrant in his fifties, is now a revered engineering professor at the University of Minnesota. After 9/11, when much of official Washington was concerned about all manner of terrorist attack, Amin was charged with overseeing research into critical infrastructure security and grid operations at the Electric Power Research Institute, a leading Palo Alto–based think tank working on the future of the grid. There, Amin and his team drew from a dizzying array of breakthroughs in fields as diverse as nonlinear dynamical systems, artificial intelligence, game theory, and network theory to develop key concepts for a self-monitoring, self-healing, self-repairing grid.

Amin and his fellow researchers identified three by-now familiar design principles of this grid of the future. The first is real-time monitoring and reaction. In much the same way as economists like Andy Haldane call for the collection of better real-time data about the performance of the global economic network, the grid needs vastly more sensors deployed at every level of its organization, from the core to the periphery. The second principle is anticipation. Current SCADA systems assess isolated bits of information with a thirty-second delay, analogous to driving a car by looking in the rearview mirror. Better monitoring and anticipation would create what Amin refers to, only half jokingly, as "self-consciousness" in the system: Advanced controllers would perform like a master chess player, modeling supply and demand several steps ahead in a chained link of events. The third design principle is isolation, or decoupling. At the first sign of failure, the grid would break itself down into islands, or isolated entities. Each island would be vulnerable to complete failure, certainly, but the systemwide cascading failures that ultimately led to the blackout of 2003 could be avoided.

Today, as these principles are being implemented, the power grid is undergoing a revolution. Attracted by a financial opportunity that may dwarf that of the Internet, an entire industry has emerged to implement these ideas, at every scale and level of organization, from the individual light switch to the grid as a whole. According to an analysis done by the networking firm Cisco, the resulting technology platform will be one hundred to one thousand times larger than the U.S. domestic Internet, with many trillions of nodes and sensors embedded throughout and millions of low-level software systems to connect them to one another.

Together, these embedded sensors and intelligence systems will imbue the grid with a sense of *proprioception*. Not quite Amin's self-awareness, "proprioception" refers specifically to "one's own" perception, or to the positioning of one's own body in space. Try a little experiment: Shut your eyes and bring both hands above your head. While keeping the fingers of your left hand totally still, use the index fingertip of your right hand to quickly touch your nose. Then, while keeping your eyes closed, use the same right index finger to touch the thumb of your left hand.

For many of us, mastering such a simple task will take a few tries, but we eventually get there—but how? Although we get vital information about our bodies in relationship to space from our sense of sight, we also have a sense of our body's positioning—where our legs and arms are, for example, at any given moment—even without looking. Special receptors, called *proprioceptors* or *stretch receptors*, are located throughout our bodies and continually relay positional information back to our brains. Our brains, in turn, take stock of the incoming data, reconcile it with what our eyes and other sensory organs tell us, and create a composite, internal sensation of our own body's orientation in relation to the space around us, even if we can't physically see our limbs and torso at the moment.

The sensors Amin envisions distributed throughout the grid will go a long way toward imbuing it with an analogous sense. Integrating the signals from these sensors will require something else, however: new protocols for encapsulating and exchanging information inside the system.

As the issues with SCADA systems illustrate, the resilience of systems like the electrical grid is highly sensitive to the design of the information protocols at their core. Protocols are the lingua franca of systems—they define how information is exchanged among their constituent parts. If the underlying protocol is inflexible, or tied too closely to a particular set of hardware and software, it can quickly become obsolete, as the underlying technology ages and is replaced. That's exactly what's happened with SCADA systems; they embody command-and-control principles from the mainframe era.

On the other hand, the Internet's protocol, called TCP/IP, shows how to do it at least partially right. This packet-focused protocol, which we described in the last chapter, has a simple, ingenious, and layered design. To simplify things, imagine an hourglass shape: At the bottom of the hourglass is an ever-expanding menagerie of hardware, from iPods to robots to who-knows-what-next, which connect to the Internet. At the top is a mirror-image, ever-expanding menagerie of software applications, from mobile apps to the future descendant of Twitter, which also connect to the network. In between is the skinny waist of the hourglass: a simple protocol—changing rarely, if ever—for encapsulating information for transport across the network. Write a new application, or create a new Internet-enabled appliance, and if it can speak TCP/IP, it can talk to anything else on the network.

In this way, TCP/IP is like the alphabet—a series of basic forms that can be used to give expression to and make connections between a countless number of ideas, but which itself changes only glacially. When the protocol's authors first invented TCP/IP, most could scarcely imagine a world of iPads and Facebook profiles, but the genius of their system is that they didn't have to—they simply had to ensure that whatever was invented could speak the universal language, and that the universal language could, in turn, transmit whatever they might like to say to one another. (A SCADA system, by analogy, fixes the relationships between the letters, enabling a more limited palette of information and ideas to be exchanged.)

FROM MACRO TO MICRO:
SCALE AND SWARM

Embedding sensor networks and the right kinds of extensible protocols into the grid will update an infrastructure originally conceived in the nineteenth century and designed for the twentieth with the management tools of the twenty-first. But the real resilience revolution will arrive when new structural features are added to the grid. The first of these innovations are microgrids—tiny, autonomous, self-sufficient, distributed systems that use locally available energy sources to power a small footprint of a few buildings, homes, or factories.

When the North American power grid was first engineered, the process of making electricity was dirty, dangerous, unsightly, and expensive, and so electricity was generated in large, centralized power plants and shipped to customers who were, at first, dozens, then hundreds, and then thousands of miles away. This is the equivalent of the mainframe era in computing: We all share the resource of the power plant, in much the same way that early computer programmers shared the resources of a room-sized computer.

To drive costs down, and in order to ensure a ready supply of fuel, power plants typically need to be situated near specific kinds of infrastructure, like railroad tracks and rivers. And for both safety and aesthetics, plants ideally need to be some distance from the people who use the energy they produce. As a result, today it's virtually impossible to look at a plug in your home and know where its electricity comes from.

In a grid complemented with microgrids, this situation is inverted. At least some of the energy on the grid is efficiently produced locally, using small-scale wind, solar, and hydropower, for example, and therefore doesn't need be shipped over long distances. In principal, its users need never be connected to a system in which distant failures can adversely affect them.

Such energy would need to be produced cleanly and safely—after all, nobody wants to live with a coal power plant in their backyard, particularly when something goes wrong. And of course the result would have to generate power cheaply—it's not enough to build a grid that delivers all the above but is exorbitant to operate.

If these cost and cleanliness problems can be solved, however, an autonomous, small-scale grid could power every city block, every home, and perhaps every person individually. The existing grid would be atomized—infinitely replicated in miniature.

No one is more interested in the concept of microgrids than the U.S. military, and it's easy to understand why. In the swarming, networked world of Arquilla's outpost and outreach, there is no front line for troops to be behind—the front line around a forward operating base extends 360 degrees. And yet these bases must be continuously supplied with fuel and matériel, typically via convoys that have to snake their way through extremely hostile territory. These lightly protected caravans of transport vehicles are slow moving and vulnerable: A 2009 study found that U.S. military forces in Afghanistan incur one casualty for every twenty-four fuel resupply convoys. The fully burdened cost of gasoline for bases in-country is a whopping four hundred dollars a gallon; at the peak of the conflict, the Marines ran through 200,000 gallons *a day*.

If their bases could be made self-sufficient, fighting forces would be able to deploy much faster and retain greater autonomy once they had done so. Under an initiative called Net Zero, the U.S. Army is focusing heavily on microgrids that use a wide variety of wind, solar, fuel cells, and other renewable sources to make this possible. Marines in Iraq and Afghanistan are already experimenting with Power Shade, a large tarp covered with flexible solar panels that fits over a standard tent and can quietly power its lighting system without the buzz of a generator that might alert enemy insurgents, and GREENS, or Ground Renewable Expeditionary ENergy System, a larger solar system that can power a platoon's command center. One proposal even suggested that the U.S. forces should encourage Afghan farmers to switch from growing poppies to growing biofuel crops—cutting down on the drug trade, building an indigenous new industry, locally supplying U.S. forces, and undermining the Taliban all at the same time.

In the civilian economy, investors and innovators are betting on microgrids as a major component of the future of energy production—companies like GE and IBM already predict that more than half of American homes will generate some of their own power in the next few decades. As envisioned, microgrids will do more than just distribute power generation: They normalize the relationship that energy consumers have with energy producers. Right now, power travels just one way: from the power station to the grid to the consumer. If microgrids ever reach their apotheosis, consumers will upload their extra reserves of power just as often—if not more frequently—as they download power from some central source. An Internet of energy would, much like its namesake, be a platform for both individual production and consumption. In this swarm version of the grid, no centralized unit controls the system. Instead, semi-autonomous little units augment the greater whole: a system of systems.

At the far edge of this vision, the renowned MIT chemist Dan Nocera is working on technologies that one day may vaporize the grid altogether—and usher in an era of personal energy—by imbuing the electrical system with a capacity even microgrids lack: storage.

In summer of 2008, Nocera set the energy and chemistry worlds abuzz with a startling announcement: His research team had achieved partial artificial photosynthesis—making fuel directly out of sunlight—with a new chemical catalyst that split water into hydrogen and oxygen. In so doing, they had solved one of the greatest challenges to the large-scale deployment of distributed solar energy: how to store the energy of the sun when it isn't shining. Connect to a box filled with this catalyst, and you can store the energy produced by even a moderately efficient solar cell during the day for intensive use at night.

"In the simplest configuration," Nocera explained to us, "a photovoltaic on your roof will generate the power you need to live when the sun shines. The surplus electricity from the photovoltaic can be fed to a small box filled with our newly discovered water-splitting catalyst to generate hydrogen and oxygen, which are stored locally. At night (or when the sun isn't out), the stored hydrogen and oxygen can be recombined in a fuel cell to give the electricity needed to power your home at night and charge the electric car in your garage while you sleep."

In Nocera's dream of personalized energy, each individual home becomes a solar power grid and a gas station. Because the system is a closed loop—sun and water to hydrogen and oxygen—very little water is actually used up. How much? Nocera's favorite measurement gauge is the pool on the MIT campus where he teaches.

"We need only one-third the amount of water in the MIT pool—an Olympic-size swimming pool—to be converted into hydrogen and oxygen every second, to solve the entire world's energy problems. A lot of new discovery, such as our catalyst, is needed to make this happen, but think of it: only one-third an Olympic-size swimming pool."

Nocera is now the head of the distributed energy storage company Sun Catalytix, where he is working to commercialize and scale his artificial photosynthesis technology. While there's much work to be done, Nocera is confident the dawn of personalized energy—and with it the beginning of genuine energy abundance—could be as little as a decade away. Already, he's formed a partnership with Tata, the Indian superconglomerate, to produce a stand-alone mini power plant that could reinvent rural electricity supply and bring power to the 3 billion people on Earth who don't have it. In such a hypothetical postgrid age, in which we don't subscribe to energy but produce it ourselves, it's not that much greater a leap to imagine a company like Walmart becoming the largest consumer energy company in the world—by selling us packets of energy, or the appliances that generate our own.

Whether such longer-term visions comes to pass, and whether Nocera's technology is ultimately the one to power them, the bottom-up microgrid approach will certainly come to complement the top-down, retrofitted smart grid infrastructure in the years to come. The two visions represent different strategies for endowing the larger electrical system with greater resilience: The latter focuses on imbuing centralized legacy systems with intelligence, the former with imbuing the periphery with greater autonomy and self-sufficiency.

Tying the two together will be newly designed protocols—"skinny waist" replacements for SCADA systems that look much more like the Internet's open TCP/IP protocol (and may indeed be some variant of TCP/IP itself, given its widespread adoption and support). These new protocols will drive greater interoperability and ensure that different local power generation technologies—including those as yet unimagined—can more easily plug and play on the grid. In this way, the smart grid merges aspects of both the energy and information infrastructures.

Embracing these approaches will substantially boost the electrical system's efficiency. The Electric Power Research Institute estimates that the fully realized smart grid will save U.S. consumers $20 billion on their annual energy bill and eliminate roughly 10 percent of national carbon emissions—akin to removing 140 million cars from the roads. Smart grid proponents refer to this reservoir of potential efficiency gains as the "fifth fuel," complementing coal, petroleum, nuclear, and renewables.

Yet the path will not be straightforward. The barriers to the adoption of the smart grid may be as much behavioral as they are technological. In early experiments bringing Amin's design principles of real-time monitoring and anticipation to the part of the grid that touches regular people's lives, in the form of residential smart meters, utilities have run headlong into an inconvenient truth: Many people simply *hate* them.

As originally envisioned, smart meters are an update on the residential electricity meters often found in the basements of people's homes—placed there precisely because they would be out of sight and out of mind. Rather than being read by a human meter reader once a month, smart meters continuously and wirelessly report their energy usage to the customers, allowing them to see their energy use patterns in real time. They also allow the electrical utility to dynamically change the price for electricity—to raise rates when demand is high and lower it when demand is low, pricing signals that are intended to encourage people to change their behavior.

In theory, this signaling should allow customers to trade changes in behavior for lower electrical bills, and it should help utilities better manage supply and demand, smoothing peaks and valleys. A more stable grid is, in principle, not only more efficient and more cost effective for consumers, it's one onto which utilities can more reliably plug in more renewable sources, lowering carbon emissions—a huge potential win-win.

But aligning supply and demand requires people to do something new: to start carefully attending to *when* they use electricity by making choices about at what time, say, their refrigerator de-ices, or time shifting to nonpeak hours when they do a load of laundry. A less variable load on the grid requires more variability among consumers—by getting some of them to use electricity countercyclically. The combination of both financial incentives and penalties designed to encourage this, combined with the real-time nature of the system, have made some early customers of smart meters feel like they're being constantly watched by Big Brother and charged for making incorrect decisions.

Unfortunately, in some rollouts of the first generation of smart meters, utilities did a nonexistent-to-terrible job of explaining the system, spurring vocal opposition groups to form that challenged the meters on grounds ranging from privacy and cost to health and safety. "I used to think PG&E was just incompetent," said one early and involuntary recipient of Pacific Gas and Electric's smart meters, after the company decided to roll them out concurrently with a price increase. "Now I think they're incompetent and *spying on me.*"

To counter such hostile reactions, many utilities are turning not to additional technology, but to a mix of behavioral economics, clear communications, and, of all things, the lowly smiley face. They're doing so by embracing a consumer-focused energy conservation platform developed by a company called Opower.

While searching for a business plan for their new energy-focused start-up, Opower's founders, Dan Yates and Alex Laskey, realized that even if electricity consumers were interested in conserving energy (something most say they support), the task was usually rendered impossible by the indecipherable, poorly targeted cookie-cutter bills they received from their utility. Here was a classic example of a poorly designed feedback loop—bills filled with irrelevant, nonactionable, and untimely information that gave consumers little sense of how their behaviors affected the system and no incentives for change.

Then, in a chance meeting, Yates and Laskey were introduced to the work of behavioral scientist Robert Cialdini, a renowned emeritus marketing professor from Arizona State University, who had spent a thirty-year career studying the science of persuasion, in particular the role of social norms in getting people to change their behavior. Cialdini had authored hugely popular books on persuasion and marketing and enhanced his scholarly research with covert fieldwork, enlisting for stints as a used car salesman, telemarketer, and other jobs that depended on persuading strangers.

One of Cialdini's central areas of research was *social proof*—the concept that people are influenced more by their immediate neighbors' behavior than by any other form of reward.

In one classic study of the power of social proof, Cialdini and his colleagues conducted an experiment with the ubiquitous towel reuse programs found in hotels, where guests are asked to keep their towels from being laundered in order to conserve water. In the bathrooms of a hotel in the American Southwest, they randomly placed three variations of the standard towel reuse sign, one focused on environmental protection, one focused on cooperation with the hotel, and a third focused on the behavior of other guests:

HELP SAVE THE ENVIRONMENT

You can show your respect for nature and help save the environment by reusing your towels during your stay.

PARTNER WITH US TO HELP SAVE THE ENVIRONMENT

In exchange for your participation in this program, we at the hotel will donate a percentage of the energy savings to a nonprofit environmental protection organization. The environment deserves our combined efforts. You can join us by reusing your towels during your stay.

JOIN YOUR FELLOW GUESTS IN HELPING TO SAVE THE ENVIRONMENT

Almost 75% of guests who are asked to participate in our new resource savings program do help by using their towels more than once. You can join your fellow guests to help save the environment by reusing your towels during your stay.

The first and second signs generated almost identical responses—38 percent and 36 percent compliance rates, respectively—but the last was significantly more effective: Almost one in two of the guests (48 percent) reused their towels. (To ensure consistency, Cialdini's team checked only rooms with single guests, and only on the first night of their stay.) Here was concrete proof that, no matter what people said they believed in, actual conservation behavior was strongly based on one's beliefs about what others around you were doing.

Based on the success of studies like this one, Cialdini, who became Opower's chief scientist, agreed to help Yates and Laskey build a system that uses sophisticated data analytics and clear, usable communications, informed by social proof, to encourage consumers to conserve electricity. To do that, the company first collects vast datasets on consumer electricity usage from the utility, combines them with data about the weather and other events, and then applies sophisticated algorithms that disaggregate the portion of a consumer's electrical usage that is related to heating and cooling, one-time events, and the running of appliances like refrigerators that are on all the time.

Then Opower uses this data to deliver a customized energy report, written in plain English, to the utility's customers that gives them a sense of how they're doing—not in the abstract, but in comparison to a tiny pool of about one hundred of their closest neighbors living in similarly sized homes, along with some concrete, customized steps they can do to improve.

Customers who are using energy more efficiently than the composite average of their neighbors are rewarded with a smiley face on their report; those in the top 20 percent are rewarded with two smileys.

And that's it. No bonus offers. No financial incentives. Just the right information, modest incentives, concrete recommendations, lots of follow-up . . . and smiley faces.

The social reward centers of the brain are so exquisitely sensitive to social norms that that's all it takes. Opower claims that 85 percent of households which receive its reports, and the various communications which follow them, modestly cut their power consumption—evidence that the normative behavioral approach works across age, income levels, and level of education. Independently, utilities that have already deployed the system report initial average energy savings from 2 to 3 percent among their customers—at a tiny fraction of the cost of doing things like retrofitting buildings or changing light bulbs. At their current scale (30 million households and rising, as of this writing), the company claims it will soon help customers save more energy than is produced by the entire U.S. solar industry.

The larger point is this: For all of the capacities that a new swarming, dynamic, smart, and resilient grid enables, people are still the linchpin to its long-term resilience. A grid that does not include, em-power, and properly motivate them to participate is hardly smart at all.

Several themes emerge from this discussion of resilience dynamics in action. Build a system with modular component parts, arranged in net-works and connected by open, "skinny waist" protocols; imbue the parts with distributed intelligence; and give people the right information and incentives, and you can create the circumstances in which resilience can flourish. The system can begin to sense its own state, sense the state of its environment, anticipate disruption, dynamically modulate its scale in response, and localize or decouple its operations when needed.

Such systems embrace an ethic of decentralization and shared con-trol—swarming—so that no single entity is absolutely in charge. But neither are they utterly anarchic. They do not do away with all central-ized authority, but instead balance it with the right kinds of local em-powerment and self-sufficiency.

Nor do these systems achieve their equilibrium in a maximally dis-persed fog of connections—in fact, quite the opposite. As we'll see next, the pattern that emerges in many resilient systems is a kind of distrib-uted, diverse density that finds its closest analogy in the city, the reef, and the garden. This is the principle of clustering.

3

THE POWER
OF CLUSTERS

In the preceding pages, we've seen how arranging a system as a network of decentralized, self-coordinating parts can bolster its resilience. But if this distributedness is such an effective strategy, why don't we see this pattern everywhere? Why, instead, do we see so many examples of just the opposite—of clustering? If the Internet, for example, makes it possible for people to collaborate all over the world, why do so many technology entrepreneurs still flock to Silicon Valley? Why do so many artists make their homes in cities like Berlin or New York? In many situations, it seems like density, not distribution, is being selected for.

Such density takes many forms and is expressed at many different scales. We see it, for example, in the powerful, almost gravitational attraction of cities and talent hubs all over the planet. In 2008, the United Nations reported that for the first time in recorded history, humanity became a predominantly urban species, with more people living in cities than in any other kind of habitat; by 2030, the planet's ranks of city dwellers are projected to swell to almost 5 billion—as many people as were on the entire planet as recently as 1987. Much of that growth is concentrated in Africa and Asia—and not just in gleaming skyscrapers, but in densely packed favelas and informal squatter settlements.

While the rise of vast megacities in the Global South has captured the imaginations and concerns of urban planners, development experts, and businesspeople alike, this urbanization is happening on smaller scales, too. The suburbs of the United States, for example, are experiencing a major reurbanization, as people abandon eight-thousand-square-foot McMansions for smaller, more densely packed living quarters nearer town centers—a trend accelerated by the rapidly growing population of older, less mobile citizens fleeing the burdens of increased energy and transportation costs.

This wave of urbanization concentrates not just people but ideas, skills, and industries, all over the world. Want to make movies in Africa? You're probably headed to Lagos, Nigeria. Want to develop pharmaceuticals in Asia? You're probably going to be interviewed in Shanghai. Want to launch a biomedical start-up in the United States? It's likely you'll at least make a pit stop along Massachusetts's Route 128 corridor.

Is all this density a good thing or a bad thing? Is denser better? Or does it heighten fragility? New research is uncovering the answers and revealing the important role such clustering plays in growth, collapse, and resilience—not only of cities, but of many different kinds of organizations and organisms. It's an unconventional investigation, being led by an unconventional kind of scientist: an urban physicist.

OF MICE AND MUNICIPALITIES

Geoffrey West, born in Somerset, England, in 1940, is reed thin with a distinguished salt and pepper beard. He is often seen wearing a breezy blue striped shirt and white pants—the quintessential sartorial choice of the summering Englishman—but his belt buckle, embedded with small bits of turquoise, reveals a different story. A theoretical physicist by training, West has spent the lion's share of his career in New Mexico, first working on projects with the Los Alamos National Laboratory and then as a researcher, board member, and finally president of the Santa Fe Institute, a nonprofit devoted to complexity research.

In the late 1990s, while doing research on subatomic particles at Los Alamos, West started to become interested in the scaling properties of biological systems. He wondered: Could the physicist's tool kit reveal universal laws of growth and scale that applied not just to infinitesimally tiny quarks and gluons, but to a mouse, a human being, or an elephant? Answering such a question would require developing a deep understanding of one of the most complex processes in all of life, and one that we touched on briefly in the last chapter: metabolism.

Though it's a term bandied about by countless diet gurus and exercise fanatics, your metabolism is not a single thing. Rather, metabolism is a complex series of chemical interactions within the body, often divided into two categories: catabolic processes that break down organic matter—like fat and carbohydrates—to make energy; and anabolic processes that use this energy to construct components of cells such as proteins and nucleic acids. This complex process—often oversimplified to the adage "calories in, calories out"—occurs in all biological organisms, negotiated throughout dense network structures that sustain life at all scales (circulatory, respiratory, renal, neural networks, etc.). Though a plant, for instance, does not have the same superficial design as an elephant, both organisms use a similar network structure to transport nutrients in and to carry wastes away.

West, along with his team of collaborators, including ecologists Jim Brown and Brian Enquist, then at the University of New Mexico, set about creating precise mathematical models of all of these biological networks, convinced that they might lead to a universal understanding of the way biological organisms scale up.

Developing a model comprehensive enough to capture the dizzying variety of all living things was no simple undertaking. West recalls years of meeting with Brown, Enquist, and the other researchers, forming a bridge between the disciplines of theoretical physics and ecology—strange bedfellows in any research endeavor, much less an effort with such an ambitious goal. By the late 1990s, however, their hard work yielded an extraordinary insight: a mathematical model that showed how every single organism in all of biology scales up in a predictable and systematic way.

What does this mean exactly? Consider a mouse. It's a relatively small organism and, for that reason, it has a fairly high metabolism and a shorter lifespan. Now think of a human being. We have a slower metabolism than a mouse and a much greater longevity: If one of our Paleolithic ancestors made it to the age of fifteen, without the aid of our advanced health care, she might live "naturally" into her midfifties. Now choose one of our planet's largest organisms—say, an elephant. As one might expect, the elephant has an even slower metabolism and an even longer lifespan. The takeaway is that, in biology, things slow down as they get bigger.

All of this was well established when West and his colleagues took up their research. What West, Brown, and Enquist discovered, however, was that all organisms in biology scale logarithmically—by powers of ten—based on the rate of their metabolism.

"The mathematical models reveal that the pace of life gets slower and slower systematically: An elephant's heart beats much slower than our heart and our heart beats much more slowly than a mouse," West told us. "Concomitantly, a mouse does not live very long; we live longer and an elephant or a whale lives even longer. If you use all this to ask about how systems grow, you discover that growth behaves in what we call a sigmoidal fashion, meaning that you start growing quickly and then your growth slows, and then you stop growing when you reach maturity. This all comes out of the theory in a very elegant way."

West and his collaborators showed that, like natural selection and genetic inheritance, these general scaling laws underwrite all forms of life.

"Scaling laws say that even though the whale is in the ocean, and the giraffe has a long neck and we walk on two legs, these are actually superficialities of design," West told us. "We are all actually representations of the same abstract thing, just scaled in a very specific nonlinear fashion. It's truly wonderful. Out of this theory, you can predict all kinds of things. If you perversely wanted to know the blood flow rate in the ninth branch of the hippopotamus's circulatory system, the model could predict it."

Based on the success of this first line of inquiry, West started to wonder if these elegant universal laws of scaling extended beyond biological organisms to social organizations like cities and companies. Is New York City really just a great big whale? Is Google an elephant?

To explore these questions, West joined a second, ambitious investigation into the systems that govern cities all over the world. Partnering with such collaborators as fellow physicist Luis Bettencourt and statistician José Lobo, the research team meticulously analyzed millions of data points from cities of every kind: lengths of roadways, lengths of electrical lines, number of gas stations, etc.

Again, the team saw unmistakable signs of universal laws at work. "The bigger you are, the proportionally fewer additional gas stations you need, fewer roads, fewer power lines, but all to the same degree. That was the extraordinary part. It was systematic. It didn't appear to matter whether we were calculating infrastructure in Japan or Europe or the United States."

The real insight came when the team correlated these factors with a dizzying array of social, cultural, and economic data: wages, number of patents produced, number of research institutions, number of AIDS cases.

Again, the researchers found universal scaling laws—but with a crucial difference from their biological cousins. In the biological world, scale made organisms slower; in the realm of cities, it made them *faster*. The bigger the city, the higher the wages were for the residents, the more patents produced there but also the greater the number of violent crimes, the more traffic, etc.

"When you double the size of the city, you produce, on average, fifteen percent higher wages, fifteen percent more fancy restaurants, but also fifteen percent more AIDS cases, and fifteen percent more violent crime. *Everything* scales up by fifteen percent when you double the size."

This scaling law proved true in every city they measured—whether that city had evolved in the thirteenth century, the nineteenth century, or at the dawn of civilization. Cities that had absolutely nothing in common culturally, politically, or geographically now seemed to share an identical mathematical scaffolding of scale.

The implications are far-reaching. Unlike purely biological systems, cities are marked by increasing returns from scale—in other words, the bigger a city is, the more it delivers per capita. (This is also referred to as *superlinear scaling*.) Cities have been selected at the population level because they get more efficient—predictably efficient—as they get bigger and faster: Life in New York is predictably faster than life in Brussels, which is predictably faster than life in San Juan. This helps explain the inexorable allure of cities: They pull us in, and as we add our number to their ranks, we in turn make them larger, more efficient, faster, and even more attractive to others.

Ah—but there's the rub. Biology's sigmoidal growth insures that an organism will stop growing when it reaches maturity. (If it didn't, our world would be filled with animals that dwarf the dinosaurs and creep along for a thousand years.) The growth of cities, however—and superlinear growth in general—doesn't stop. Not only that, cities grow exponentially, meaning you get some stunning returns on capital creation, but you also have plenty more people to contract disease and create pollution.

Worryingly, West and his colleagues found that, absent any adaptation, systems that follow a single exponential growth curve inevitably collapse. If you live by a single curve—reaping the benefits of a single mode of wealth and capital creation—you can also die an ignominious death by the same single curve. Like the robust-yet-fragile financial markets before the 2008 crash, cities that become overly reliant on just a few forms of value creation can find themselves enjoying a golden age followed by catastrophic decline. (Think Detroit.)

That is, unless you innovate into some new state and restart the clock. Like a surfer, this process requires shifting your board from the wave of one innovation growth curve to catch the even bigger rising wave on the horizon.

"The only way to avoid the collapse is to invent the steam engine, electrical lights, computing, the Internet," West told us. "The kicker is that, as the pace of life speeds up, the pace of innovation needs to speed up as well. This wristwatch of mine is a big fake. We are not in linear time. Growth and innovation are going faster and faster."

How do successful cities continue to innovate, reinvent themselves, and leap from one curve to the next, even as the waves come larger and crest faster? The answer is not in their scale, but in their clustering of density and diversity.

Cities are large piles of small, very different things—neighborhoods, networks, innovators, and infrastructure. These tie together different people and groups in loose, informal, ever-shifting affiliations. Think of the more famous passages from storied city dweller Jane Jacobs's writings, in which the streetscapes of the West Village form a vibrant collage of densely packed interactions, each embodying a genuine diversity: the local news vendor selling papers; the cop on the beat; the commuters coming and going—each represents a complex layer of differing scales and purposes, commingled with one another.

"One of the great things about being in a city is that there are a lot of crazy people around," West added with his characteristic dry humor. "I suppose that's another way of saying cities have lots of cognitive diversity. Some of them are the dregs but some are not. They provide a landscape that allows the spectrum of ideas to blossom. As the city grows, this makes it more and more multidimensional. Cities seem to open up: the spectrum of functionalities, job opportunities, connections, etc. That is key to the vitality and the buzz of successful cities."

It's this densely packed distributed diversity that lends cities, like the coral reefs we've explored earlier, their ability to innovate when one economic or industrial wave crests—it ensures there are always new groups jockeying to embrace the next wave, and different kinds of thinking and capacities are at the ready to deal with the inevitable disruptions that ensue.

This pattern of nested, dense-but-diverse clustering has application in domains well beyond traditional urban environments and modern business. On the other side of the planet, in the remote forests of Indonesia, one passionate conservationist is embracing these design principles in a quest to save the country's disappearing biodiversity, restore its natural environment, and bolster the human society—and economy—that must live cheek by jowl with both.

CITIES IN THE JUNGLE

It's a balmy fall day on the Upper East Side of Manhattan, but Willie Smits has been sitting in a dimly lit living room, perched on the edge of a rattan chair, for the bulk of the afternoon. Smits, fifty-four, a Dutch-born forestry expert and orangutan conservationist, is speaking to a group of young ecologists and activists who have come down from Boston to meet with him.

A relentless schedule has taken him, in a matter of days, from Singapore to Amsterdam to Denver to Dallas and then back around to New York City, and his jet lag leaves him looking visibly exhausted. Yet despite his obvious fatigue, Smits can only listen for brief periods of time before launching into bursts of exposition. He offers one detail after another regarding reforestation in Indonesian Borneo—the place he has called home for some thirty years. His voice is measured, his delivery encyclopedic. But when the conversation shifts to focus on illegal wildlife traders and their cruelty to the orangutans, his tone sharpens and he becomes agitated.

"We must get those bad guys."

The words, almost childlike, spoken in his thick Dutch accent, have a piercing, palpable anger about them. The room grows silent.

"I've tried to get them through conservation; I've tried to get them through government; I've tried to get them through the legal system; I've tried all of it. The only thing I have left now is business. We have to make more money doing the right thing than they can make destroying the planet."

This is what visitors have come to hear about. They settle back into their seats as he launches into a description of his vision—brilliant, complex, and conceived in a mind almost absurdly mechanistic.

"It is all about integration," he begins.

Smits has been an animal lover his entire life. As a teenager, he used to clean up at local checkers tournaments and use his prize winnings to fund homemade nature films. But he did not initially set out to save the orangutans. After studying tropical forestry and microbiology in his native Netherlands, he traveled to Borneo in the 1980s; there, he fell in love with the remote, tropical landscape and he set down roots to begin a life of scientific research in the rain forest.

Borneo—more than half the size of Alaska—is the world's third largest island. It is divided into three territories: Malaysia, Indonesia, and the small nation of Brunei. With fifteen thousand plant species, more than two hundred varieties of mammals, and hundreds of species of birds, amphibians, and freshwater fish, the island, like the Great Barrier Reef, is a biodiversity hotspot. It proved a wonderland for a young research scientist like Smits. "Oh, I was very happy back then with my microscope and my eureka moments," he told us.

One of these eureka moments was a discovery about the role that certain kinds of fungi, mycorrhizae, play in the regeneration of the tropical rain forest. Smits naturally assumed the next step would be to begin the long process of publishing his research in a peer-reviewed academic journal, but in 1989, a chance encounter changed the entire direction of his life and career.

"I was walking through the market when I heard the sound of

gasping—dying, really. I turned around to try to see where such an awful sound was coming from and someone shoved a cage in my face. There, before me, were the saddest eyes I had ever seen."

Smits came face to face with a dying baby orangutan being sold in a cage in the street markets of Balikpapan, in eastern Indonesia. Though they had locked eyes for only a few seconds, for the rest of the day he couldn't get the animal out of his mind. Later that night he returned to the market and found the baby, barely breathing, tossed atop a garbage heap.

"They managed to salvage the cage," he noted acidly.

Smits folded the baby orangutan into his arms and broke into a run. The market vendors chased him, demanding money for the dying baby, but he and the infant managed to escape.

Back home, holding the tiny baby that he later named Uce, Smits experienced what can only be referred to as a calling. He nursed baby Uce back to health just like a human baby, holding her and giving her warmth. He promised to keep her and her family safe from the illegal traders who make a lucrative profit capturing orangutans in the wild and selling them off as caged status symbols in both national and international black markets. The promise instantly transformed Smits from a research scientist into a crusader—one with a nearly impossible challenge.

In prehistoric times, orangutans roamed freely throughout Asia, but as human beings came to dominate the planet, our increasing footprint pushed the apes into the few remaining areas of untouched rain forest in Sumatra and Borneo. In the last fifty years, logging, massive annual forest fires, and the wildlife trade have jeopardized their very existence. Now, orangutans—one of the four great apes, including chimpanzees, bonobos, and gorillas—sit near the top of the list of the world's most endangered species. In nearby Sumatra, their numbers have been halved since 1993; only around 6,500 remain there today. In neighboring Borneo, approximately 50,000 orangutans are left. A UN report released in 2007 estimated that the tropical rain forests in both places

were being cleared so rapidly that, without urgent action, in ten years 98 percent will be gone.

Contributing to the orangutans' fragility is the gentle pace of their individual growth and development. Just like humans, baby orangutans cannot survive in the wild alone without the constant care of their mothers. Orangutan mothers are known for nursing their infants longer than any other mammal on Earth (approximately six or seven years), and the orangutan birthrate (one baby every six to seven years) is one of the lowest on the planet. All of this is due to the lengthy education each infant must receive from its mother to survive in the jungle, learning to decipher the hundreds of different types of edible fruit, bark, and leaves. As the forests are increasingly clear-cut for agriculture, many of the orangutans have moved onto the farmland looking for food; once there, they are often captured or killed by the timber and plantation workers. Orphaned infant orangutans are kept as chained and caged-up pets or traded for lucrative sums on the international black market, where Smits first found Uce.

Back on the Upper East Side, Smits's meeting with the activists from Boston is winding down. Orangutan figurines of finely carved wood and brass line the mantel of the Manhattan living room—his host is a devoted conservationist—and they serve as talismans for the battle against extinction that is constantly lurking on the horizon. But Smits is clearly frustrated. He doesn't feel as though he has fully articulated his vision for saving the orangutans. There never seems to be enough time to fully explain how all the pieces fit together, and some of his visitors still look confused. He grows frustrated and starts near the beginning again.

When Smits first nursed Uce back to health, he wanted to create a temporary home for her and others like her until they could be returned safely to the forest. The orphaned infants needed constant care and supervision while the older orangutans—often physically and mentally abused by their owners—required intensive rehabilitation. So in 1991, Smits started the Borneo Orangutan Survival Foundation (BOS), dedicated to the conservation of orangutans and their habitat.

As word of his efforts spread, Smits began to receive other endangered species—not just orangutans—left homeless in the wake of the forest razing. What started out as a temporary way station for orangutans gradually developed into a network of permanent rehabilitation centers for all of the region's endangered species, a Noah's ark in the tropics. Over the last fifteen years, BOS has taken in close to two thousand animals for rehabilitation; seven hundred of those are now back in the wild.

Yet, for all these successes, over the last decade, Smits has seen little progress toward rain forest conservation—the underlying reason for the animals' crisis. If anything, the rampant destruction has accelerated. It started to become clear that orangutans were only one piece in a much larger puzzle that needed to be solved. Though the orangutans were able to adjust, albeit imperfectly, to the increasing human footprint over the last century, nothing prepared them for the recent accelerated annihilation of the forests across Southeast Asia. The cause of the rampant deforestation destroying their home can be explained in two words: palm oil.

One of the world's leading agricultural commodities, palm oil is used in 50 percent of all consumer goods—everything from soaps and detergents to breakfast cereals and vegetable oil—making it nearly impossible to avoid on the supermarket shelves. If you are reading this book in the United States, it's all but certain that something you'll eat today was cooked with it.

The oil is extracted from the pulp of the red fruit of the *Elaeis guineensis*, the oil palm tree. Grown in monocultural plantations across Southeast Asia, it requires significant cropland for cultivation, which has in turn driven vast deforestation: Between 1967 and 2000, land designated for oil palm plantations in Indonesia skyrocketed more than 1,500 percent, from less than 2,000 square kilometers (770 square miles) to more than 30,000 square kilometers today.

And the trend is accelerating: Spurred by government subsidies over the last few years, European energy companies have started designing

generators powered by biofuel derived from palm oil. These biofuels have been falsely touted as a clean alternative to fossil fuels; rising global demand for them has been the motivating force behind an even faster clearing of more huge tracts of Southeast Asian rain forest, along with the overuse of chemical fertilizer.

Even worse, space for the expanding palm plantations is often created by draining and burning peat land, which, in turn, sends huge amounts of carbon emissions into the atmosphere and creates devastating forest fires that can burn for months at a time. New research estimates that 2.5 billion metric tons of CO_2 were released into the atmosphere as a result of forest fires in 1997 and 1998. Today, Indonesia ranks behind only the United States and China in terms of total man-made greenhouse gas emissions.

Smits realized that the goal of orangutan conservation was too narrow to address the systemic crisis he was experiencing all around him. It no longer seemed advantageous to try to save one species in isolation. Considering the highly networked nature of the tropical rain forest ecosystem, it could even be detrimental.

"I don't want to do symptom treatment anymore," he told us. "I'm trying to get to the heart of the problem."

To make good on his initial promise to Uce, Smits started to think about ways to intervene in the ecosystem as a whole: engaging with the people, the animals, and the ecosystem as three equally deserving interdependent components of a tightly coupled system.

When he broached this idea with some traditional players in the international conservation community, however, the responses were often chilly. Many of these organizations' funding channels had been optimized to focus on one charismatic "brand" species. The orangutan people looked out for the orangutans; the leopard people looked after the leopards. Focusing on one single sexy species was a powerful fundraising tool, even if it was an ineffective conservation tool, and the groups tacitly agreed not to get in one another's way. Taking on the system as a whole meant upsetting these unspoken agreements.

But that wasn't the only impediment to collaboration. Many long-standing Western conservation organizations had galvanized their supporters around the far cheaper idea of preserving rather than regrowing forest: buying up tracts of land that would be protected from the intrusions of humans and industry.

Creating such reserves is a classic approach to risk mitigation, and it's far less expensive than regrowing forest from scratch on decimated land. Yet while such mitigation efforts are critical to slowing the degradation of the forests and buying the overall system time, they are not without their problems: If the needs of the people living in and along these tracts are neglected, the political will to maintain preserves can falter over time. In addition, the animals these reserves were supposed to contain often migrate beyond their borders. In areas of Kalimantan in Indonesian Borneo, the phrase "*Mana yang lebih penting menyelamat-kin orangutan atau kami?*" (What is more important to save: orangutans or us?) was an oft-heard refrain in meetings between conservationists and frustrated residents, for whom conservation had become a zero-sum game: "orangutans win" equals "humans lose."

While many of the more traditional conservation groups were still trying to keep the humans out, Smits came to believe that the only viable long-term solution would require human beings, human enterprise, orangutans, and myriad other species to live together, just as in the overlapping communities of a city. Long-term resilience would require the marriage of viable economic growth with ecological preservation, underwritten by economic models that were designed to encourage reforestation and biodiversity protection—the creation of adaptive capacity.

Smits became a man obsessed by a vision of systemic integration, scribbling down countless notes in journal after journal, often accompanied by engineering designs and system flow charts. But before he could even begin to implement some of these designs, he needed to start with a little gardening. Willie Smits set himself a modest task: He would regrow the rain forest.

He chose the most barren tract of land he could find. The area around Samboja village, considered the poorest district in East Kalimantan in Indonesian Borneo, had virtually no fertility left. The local residents spent 22 percent of their income just to buy water, and close to half of the population was without work. Without any buffer zone of biodiversity, they were subjected to constant floods and fires.

"If I can do this in the worst possible place I can think of, no one will have any excuse to say, '*Yeah, but* . . .' Everyone should be able to follow this."

By 2001, Smits had crafted an initial design, something of a hybrid of an urban cluster and lush garden. Instead of planting rows and rows of a monoculture crop like oil palms, he envisioned three rings of mixed forest over approximately 2,000 hectares, or 5,000 acres, of land. The outermost ring would consist of flame-retardant trees that would protect the entire preserve from forest fires. The middle ring would contain mixed forestry: the bulk of the plant diversity in tropical hardwoods and food crops like pineapple and papaya to feed the local people as well as the orangutans. And in the centermost ring, out of reach of the poachers, there would be a preserve where Uce and her family could roam free along with all of the other endangered species under Smits's charge. If he can pull it off, the regrowth of the rain forest—and subsequent economic benefits provided by the return of the ecosystem services—would present a viable alternative to the destructive, self-replicating platform of the oil palm monoculture.

He christened the project Samboja Lestari, or "Everlasting Forest."

Smits had spent most of his adult life studying the various trees and plants of the Indonesian forests, so he was uniquely prepared to take on the challenges of rain forest reconstruction. Yet this goal will be impossible without the sustained engagement of the Dayak people, who live with and within the ecosystem. Fortunately, Smits does not undertake this venture as a complete stranger: He has lived in the area for thirty some years, officially becoming an Indonesian national. His wife is a traditional Dayak tribal princess from the Indonesian island

of Sulawesi, and he raised his sons in the rain forests of East Kalimantan, speaking several of the local dialects. He is considered a part
of the community, so much so that he has been given a Dayak name
and he performs regularly with the local Dayak band on birthdays and
holidays. "None of this would be possible if I was perceived to be an
outsider," Smits said.

Smits works to retain that trust, starting with paying the local
people what they all agreed was a fair price for their land. After each
transaction, he was fastidious about making each and every purchase
of land compliant with government regulation. He wants to protect
his initiative from the corrupting forces of the timber and oil palm
mafias—especially if his reforestation plan works and valuable timber
reappears on the barren wasteland—so he legally insured that BOS, his
conservation group, would hold the rights to the land in perpetuity.

Once the area was safely and fairly secured, the planting process
could begin. In the beginning, not a few people thought he was crazy:
this large white Indonesian man paying people generous sums of
money for land known to be completely barren. When he stood on the
desolate scrub that was to become Samboja Lestari for the first time,
Smits could not hear a single sound. The tract of burned soil was so
bereft that even the insects had fled.

Understanding Smits's vision of forest regrowth requires thinking
systemically, and therefore in a variety of different time signatures.
Imagine a conductor in front of an orchestra. One wave of the right
hand indicates the start of the prelude (the first layer of forest growth),
while a later wave of the left ushers in the rising movement of the forest's full canopy, the growth of the tallest tropical hardwood trees that
will shade the entire forest and capture the humidity that is so essential
to the rain forest's rich biodiversity. Unlike a monoculture, Smits's approach layers in processes with different scales, different time signatures, different structures—all in a dense, interconnected cluster.

This type of planning, of course, imitates nature's evolutionary
design. As Amory Lovins, expert on alternative energy sources and

advocate for Smits's work, put it, "Nature has had 3.8 billion years of experience at getting multiple benefits from single expenditures: Willie's work is the extraordinary example of modeling nature and creating this same abundance by design."

Like a symphony, the first temporal movement of Samboja Lestari's growth will lay the essential foundation for the full blossoming to follow. Because the barren land had been overrun by an insidious cyanide-producing weed called *alang alang,* Smits envisions planting the outer ring—or zone—with quickly growing trees that would both protect the land against fire and shade out these weeds. His first layer includes *Acacia mangium* trees: fairly low value, but fast growing. Along with other fire-retardant trees, they would protect the inner rings and provide shade to encourage the returning microclimates.

"The reason a forest is multilayered is so it can make use of light at different elevations, store more carbon in the system, and then provide more functions," Smits said. "So, just like in nature, we will plant fast-growing trees and then follow that up with the slower-growing primary forest with a diversity that can optimally use as much of the light as possible, at different times, as it grows."

Once the acacia trees are growing and offering shade to subsequent growth, his design calls for the cultivation of the middle zone—the proverbial second movement—more than 1,000 hectares intended to contain the bulk of Samboja Lestari's biodiversity.

In temperate zones—between the Urals and England, for example—there are approximately 165 species of trees. By contrast, in this second zone, Smits envisions a half million trees belonging to more than 1,300 species. His earlier eureka moment involving fungi and tropical forest regeneration served him well: He has used his discoveries to create specially designed compost with microbiological agents made from sugar, food waste, sawdust, cow urine, and (what else?) orangutan excrement. Every tree planting will get a hearty dose of this compost to enrich the nutrient-starved soil. Smits plans to transplant approximately 150,000 saplings from his nursery. They will be integrated into the mixed forest—layered over time—as soon as the soil is enriched enough to sustain their growth.

"Nature is able to recover from almost anything because of its bio-diversity: the structural biodiversity and the species biodiversity," Smits says. "This is the underlying basis of resilience in systems. We must find a way to replicate that if we have any chance of survival."

Because the growth of these second-zone tropical hardwood trees can take anywhere from ten to fifteen years to reach full canopy, Smits intends to use the land alongside the planted saplings for food crops. Papaya, lemons, pineapples, watermelons, beans will be planted—then later, chocolate, chilis, coffee, and other crops will cycle in a new rotation of plantings. BOS will buy the farmer's surplus to feed the orangutans and other wildlife in the inner-zone preserve, bringing the farmers an additional source of income.

Once the first and second movements have been achieved, Smits will be able to focus on the innermost zone of 300 hectares: the wild-life sanctuary, animal nursery, and a forest research facility. There, the healthy orangutans will be released into the preserve along with other wildlife species. Orangutans that require isolation—those suffering from hepatitis and other diseases—will live in groups on manmade orangutan "islands."

Step inside the mind of Willie Smits and see what he sees: years and years of forest regrowth across Samboja Lestari—the "Everlast-ing Forest"—occurring in mere seconds. The fast tempo of the first movement, the acacias, met by the slower, rising tide of the tropical hardwood species. Enter the staccato rhythms of the food crops and a baseline of the micronutrients enriching the soil. Although it appears like a naturally occurring rain forest, this controlled growth is as con-sidered as an eighteenth-century French garden. Like the urban plan for New York City, it accommodates many different spheres of life in a single dense, diverse layered system—Jane Jacobs in the jungle.

Finally, it will be time for the third and final movement: the crescendo—the financial engine that will make the entire system economically viable. In order for Smits to design an alternative to the palm oil economy, he must integrate a crop into Samboja Lestari that can viably compete with palm oil's productivity. Fortunately, Smits has a powerful candidate, one he has spent the last thirty years studying: *Arenga pinnata*, commonly known as the sugar palm.

"When I married my wife, the ritual dowry to marry a young bride from her tribe was six sugar palms. I asked myself, How can I marry such a lovely girl for just six sugar palms? I started doing loads of research and I really saw the potential of this incredibly productive plant."

A. pinnata is the Swiss Army knife of palm species. Its most important and well-known product is a sweet sap called *saguer*, used as a drink and as the raw material for sugar production, but there are dozens of other uses for the tree's various parts, ranging from food to roofing materials. Better still, the plant is not flammable, so the sugar palm could bolster the ring of fire protection around Samboja Lestari. And, the pièce de résistance for any conservationist, *A. pinnata* will grow only in a mixed forest.

"Sugar palms couldn't be more different from oil palms," Smits said. "To begin with, they only grow in the degraded land of a secondary forest. And they can only survive in a polyculture. If you try to grow them in a monoculture, they turn yellow and die. This is one of the main reasons why they haven't been invested in yet. The big companies want to have big areas with total control. They want a very simple, scalable system, but sugar palms don't grow that way."

Smits has been fascinated with the multifunctioning plant ever since that first encounter in his wife's village. Its versatility combined with its low-maintenance growth—unlike an oil palm, *A. pinnata* doesn't need any fertilizers or pesticides—were reason enough to consider incorporating it into Samboja Lestari. Smits, however, had his eyes set on a bigger prize: He was convinced that, with the right processing, the sugar palm could produce ethanol—the fuel extracted from the chemical compounds of sugar. If his hunch proved correct, not only would Samboja Lestari achieve successful forest and wildlife regeneration, it could also provide a sustainable alternative energy source and provide wealth for its human inhabitants. Because, unlike the corn-based biofuels being developed here in the United States, the sugary sap of *A. pinnata* is not coupled with the food production system, so using it for fuel would not impact the food supply, the underlying cause of a crisis like the tortilla riots. The sap can be extracted from *A. pinnata* without destroying the tree. The sugar palm is basically a self-sustaining biological machine for converting and storing solar energy as sugar: It needs only rain, CO_2, and sunlight to produce more juice, all of which are available in surfeit in the tropics. In Smits's vision, a series of mixed forestry systems cultivating *A. pinnata* for biofuel could scale beyond Indonesia without compromising food security in some other unrelated corner of the globe. And all of this, in turn, could help save the forests for other endangered species, like his beloved orangutans.

But it almost goes without saying: The more beautiful the vision, the more complicated the execution. Before even beginning to integrate sugar palms into the system, Smits needed to solve a fundamental problem: the sugar sap's rapid fermentation. Because the sugary juice begins to ferment almost immediately after being exposed to the air, Smits needed a technology that could collect and process it closer to the trees—a portable sugar-processing factory.

Smits turned to the tappers of Sulawesi, a neighboring island several hundred miles from Samboja Lestari, where local men had been tapping the sugar palm for thousands of years, to learn more about how traditional sugar processing was done.

The results were a mixed bag. Twice a day, these tappers climb high up into the sugar palm trees to cut into the vessels of the plant at just the right angle. Although this process can be time-consuming, *A. pinnata* requires little additional maintenance in exchange for its productivity. In fact, the more it is tapped, the more sugary sap it produces.

Traditionally, these tappers would use firewood as the energy source to transform the resulting sugar palm juice into pure sugar. This garnered a small profit for the tappers but it also presented a whole host of new problems for the local people. Sugar processed over the fires had a limited market, as it was often poorly stored and contained contaminants from the processing such as bees and fire ash. With no standardized procedures, the quality was inconsistent from tapper to tapper. Much of the unsold sugar was then made into a moonshine the local people referred to as "the Potent Spirit"—a high-octane, inexpensive intoxicant readily accessible to the unemployed men in the village.

But the most damaging effect of existing sugar palm processing methods was the fires themselves. Wives and daughters of the tappers spent long hours searching for enough timber to make a fire big enough to process the sap. This stole time away from their families and other more potentially productive tasks. It also added to the large-scale devastation of the forests and led to chronic respiratory and vision problems for the women and girls who sat day after day in smoke-filled huts.

Based on what he learned in Sulawesi, Smits was able to design a small prototype portable sugar-processing factory that was more energy efficient, cleaner, and safer to use. Then, in 1996, Smits happened on a piece of exceptionally good luck: A power company called Petramina started operating in the nearby area of North Sulawesi. Its plants powered by geothermal heat, Petramina struggled to find a use for its waste heat, in the form of excess steam. Smits realized that he could use this steam—piped into his portable sugar-processing factory—to heat the pans used for sugar production and thus completely eliminate the need for fire and firewood. Petramina was thrilled by the offer: In exchange for supporting an ecologically beneficial effort, the detour through Smits's factory would cool their steam back into more useful liquid water.

Instead of burning wood for fuel, steam now converted the sap to sugar, saving an estimated 200,000 trees a year. With thousands of liters of sap processed in the factory every day, the cooperative could manufacture a consistent palm sugar product for export. And, best of all, by including a yeast fermentation process, the geothermal heat allowed

Smits to extract ethanol from the sugary juice. As a result, the families involved in the cooperative gained access to both fuel and electricity.

The prototype sugar-processing factory was initially so successful in North Sulawesi that Smits envisioned a mini version of it to be delivered to Samboja Lestari by helicopter in three modular pieces. Instead of geothermal steam, the minifactory—dubbed "the Village Hub"—would be powered by solar panels.

"In the tropical regions, we still have vast amounts of land, but we don't have jobs, so we have to look at simple, ecological, sustainable systems that are going to be based on what can be produced," Smits said. "There are ways to provide energy from this tropical belt where you have all this rainfall, all this land, all these people, and the right climate in terms of solar radiation. But we need technologies like the Village Hub that can function as an integrated unit."

Smits's goals may be rooted in conservation ecology and social justice, but he is hard-nosed about the economic and political realities. He's relentlessly focused on making his system profitable—including patenting the underlying technologies involved.

"I didn't want to give the hub away because every time we gave technology away, the big companies would try to squeeze as much money out of it as possible. I don't want the corporate social responsibility model anymore: Make a lot of money and just a little goes back to the local people and everyone has to get down on their knees to say thank you. I don't want to work that way. Our model guarantees that a real income for local people is built into the model."

In order to insure that this happens, Smits envisions a franchise structure. Everyone using his technology would have to agree to strict guidelines for the cooperative—protected by Dutch law—all guaranteeing that the profit returns to the local people. Each family that opts in to Samboja Lestari would commit to conserving the environment by protecting the forest's biodiversity. In exchange for their commitment, the families could then make a living from the forest's ecosystem services.

Everything about Samboja Lestari's plan is interwoven into everything else: the quintessential example of a zero waste system. So much so, it can be difficult to describe its many flows and cycles in a linear way. "The whole system is elaborately and intricately interlinked to get the most social, ecological, and economic benefits," Amory Lovins says. "I think it's the best model in the world for what we all need to be doing in our respective endeavors."

It accomplishes this by embracing the very themes we've been speaking about until this point: Like the city, the design of Samboja Lestari is a cluster, at once dense, diverse, and distributed. It can be connected to the global economy (through the global market for biofuels) but is also decoupled from it. It links immediate economic transactions that benefit the Dayak people to slower patterns that are needed to ensure the long-term viability of the orangutans and the even longer ones needed to restore the forest ecosystem.

And, according to Smits, it's scalable: he believes the three-ring repeating pattern can be replicated at many sizes, as sugar palms can thrive in any place that has more than 750 ml of rainfall and sits at an altitude of less than 200 meters. Smits imagines thousands of Samboja Lestaris linked up across Indonesia, across the tropics even—three-ringed densely mixed forest gardens sustained through the productivity of the sugar palm. "Scale is not the enemy, only the wrong kind of scale," says Smits.

The Samboja Lestari that exists in the mind of Willie Smits is breathtaking in its ambition: an integrated system design that treats human beings, other species, and the ecosystem as co-equal participants in a whole. Yet the reality on the ground is still a prototype in progress. The challenges that must be surmounted if it is ever to be more than that are daunting.

First, Smits must overcome the resistance of mitigation-focused conservationists who argue that rebuilding the forest is too expensive and does not deliver enough bang for the buck. Smits is the quintessential embodiment of an adaptationist; he has seen the wide-scale destruction of the forests across Borneo and he knows there is little time left. The car, to return to an analogy from the Introduction, is tottering on the edge of the proverbial cliff. But if Samboja Lestari can be made to work, Smits believes, it could provide the "air bags" for the system, in the form of renewable energy systems and initiatives for forest regrowth, all the while widening the basin of possibilities for the Dayak people and getting the orangutan's "parachutes" open.

Smits has also drawn criticism from some quarters for not publishing more scientific data about his efforts, leading some in the field to question the impact of what's actually been accomplished. As befits a man with an evangelist's zeal and a sense that time is short, Smits's response to these critiques has largely been to ignore them.

Other concerns cannot be so easily dismissed. For Smits and Samboja Lestari to succeed, he must navigate the extremely complex waters between private and civic interests, keep the local community committed and engaged, raise the resources to nurture the project during its years-long path to fruition, and maintain momentum through the inevitable failures and reversals that come with tackling anything this comprehensive. Doing so sometimes requires Smits to be in more than one place at the same time and sometimes calls on him to assume many roles—businessman, activist, scientist, social leader, and storyteller—that can be in conflict. He has a reputation as a brilliant, almost insanely dedicated solo operator, and his vision, unlike the emergent, cross-pollinating innovation that West describes, is entirely *centrally* planned. For Samboja Lestari to fulfill its promise it will take a whole community of equally dedicated, and equally empowered participants and supporters, who can make—and remake—it anew. And it will call upon Smits to demonstrate a very specific form of leadership, which we'll explore in greater depth in chapter 8.

Finally, while Samboja Lestari represents a daring vision, we must be careful not to fetishize it as an endpoint in itself. If it succeeds, it will not restore some sacred balance or achieve a stasis under glass that freezes the relationships between its constituents. Rather, it would restore the eroded capacity of the human community, economy, bio-diversity, and ecosystem to contend with unanticipated future disruptions. In Samboja Lestari, Smits and the Dayak people hope to find not an answer for every possible calamity, but a system that can give them a wider array of future choices when one of those calamities arrives.

Smits seemed energized by the challenge. "A lot of people are afraid of the complexity, but we have to embrace it. It can't be in a single species, block by block. That is the big mistake that modern agriculture and forestry are making. They are always chasing the biggest profit and looking for the quickest exit strategy. Well guess what? There is nowhere new to go anymore. Exit to where? We only have what we have. We need to work with it, for everyone's sake."

Like most systems, Samboja Lestari's long-term viability and resilience will begin and end with the choices and commitments people make, their individual and collective responses to change, and their ability to work together. And so it's natural to now turn to the next part of this book, in which we'll explore the roots of resilience in people, groups, organizations, and communities.

Our journey in the chapters to come is organized along increasing levels of scale. We'll begin with the underpinnings of personal resilience—how might we enhance individuals' capacity to bounce back psychologically in the face of potential trauma? Then we'll examine the contours of group collaboration in the face of disruption—how do we get people to work together when it counts? We'll also explore how to enhance the cognitive diversity of such groups—how do we ensure we consider the widest array of options? Then we'll look at how certain kinds of leaders can amplify the resilience of entire communities. We'll end by offering some thoughts about what these lessons might mean for society as a whole.

4

THE RESILIENT MIND

Thus far, we've explored the resilience of large, autonomous systems that are not driven primarily by individual actions and decisions. What happens when we look through the other end of the telescope—exploring resilience through the people and communities that live with and within those systems?

The picture that emerges, at least for individuals, is tantalizingly optimistic. Not only is personal resilience to trauma more widespread than previously believed, but new research suggests there are concrete things we can do to bolster it—to help every person better contend with inevitable disruption and difficulty and, by doing so, amplify the resilience of the larger systems and communities in which they reside.

THE LINGFIELD FOUR

In 1945, following World War II, an orphanage in the village of Lingfield in Surrey, England, arranged to take twenty-four young child survivors of the Holocaust into their care. Most of the children, between the ages of three and eight years old, were either arriving from concentration camps like Auschwitz and Terezin or had been living in hiding. They were already veterans of traumatic stress: Children from the Terezin camp had been present at mass hangings and many of them over the age of six had been forced to pass boxes of human ashes back and forth. The Auschwitz children had been surrounded by the stench of dead bodies, waking to the sight of the crematoria smoke each day. Those in hiding had been betrayed by former friends and neighbors and often forced to take different names and identities from their parents. Their lives had been so plagued by disruption and uncertainty, upon their arrival at Lingfield, one of these children asked, "Will the walls be here tomorrow?"

The four youngest children were only months old when they arrived at the Terezin concentration camp. They had spent the first two and a half years of their lives in the camp's Ward for Motherless Children, cared for by a string of prisoners waiting for deportation to Auschwitz. One of their inmate caretakers described their environment in a letter in 1946:

> There was always too much work and too few people to help me. Besides looking after the children we had to see to their clothes, etc., which took time. We looked after the bodily welfare of the children as much as possible, kept them free of vermin for 3 years, and we fed them as well as possible under the circumstances. But it was not possible to attend to their other needs. Actually we did not have the time to play with them.

These four youngest children arrived at Lingfield severely malnourished and small for their age. They had never received any consistent caretaking and, for the most part, they focused on their own group as a source of attachment and comfort.

Eventually three of these toddlers were adopted, while the fourth, a young boy referred to as Berl, returned to Lingfield after two unsuc-cessful attempts at living with new families.

In 1979, when all four of them were thirty-seven, an American psy-chologist named Sarah Moskovitz found these child survivors, origi-nally described in a monograph by Anna Freud and Sophie Dann. She conducted a series of follow-up interviews with them both in 1979 and 1984 to document their progress over time. Berl and Leah, the smallest and weakest of the youngest four, suffered the most. They both strug-gled socially and academically. Leah was referred to as "the whiner" by the adults at Lingfield and, in her interview with Moskovitz, she expressed deep feelings of shame and anxiety as well as noting that she suffered from sleeplessness as an adult. Berl had refused to stay with his two adoptive families and lived in the institution until, at age seventeen, an aunt and uncle adopted him and brought him to the United States. When Moskovitz found him, he was still living with them, a barely functional thirty-seven-year-old man debilitated by numerous severe pathologies.

Berl and Leah survived but they struggled, riddled with anxiety, shame, and sadness about the past. More surprising were the interviews with Jack and Bella, the other two members of the quartet. When Moskovitz found him, Jack was happily married with a supportive wife and two children. He owned his own taxi in London and he described the pleasure he took in meeting new people and the adventure of each ride. He struggled with depressive bouts from time to time, most often originating from his desire to know more about his mother, but, by all accounts, he was managing life well.

Perhaps most remarkable was the interview with Bella. Upon arrival at Lingfield orphanage, Bella immediately started exploring, finding her way from the dining room all the way across the institution to her room. After living there only a short time, she earned the moniker "Bella-Pick-It-Up" because she managed to coerce all the older kids to pick up her things for her. She "chose" her adoptive parents by march-ing up to her future father and sitting on his lap. When Moskovitz interviewed her as an adult, she described Bella as sunny, vital, and confident:

> Despite her husband's recent heart surgery she believed that they could come through anything together. She had started a business dealing in art which was doing well and which she enjoyed. She also worked as magistrate on cases involving children.

In spite of everything she had encountered in infancy and early childhood, Bella not only survived, she flourished. Moskovitz called her a model of resilience and encouraged mental health workers to reach out to other child survivors as a means of gaining greater insight into risk and adversity.

How could four children brought up in the exact same traumatic circumstances land in such vastly different places in life? Why do the Berls and Leahs of the world languish while the Jacks and Bellas cope and even flourish?

These questions began getting serious attention for the first time in the 1960s and early '70s when a number of psychologists working at the intersection of child psychiatry and developmental psychology began to investigate the early childhood factors that impeded healthy growth and development, including mother-child separations, divorce, prenatal complications, and, arguably the greatest risk factor of all, poverty. Much of this work was predicated on clinical psychologist Norman Garmezy's pioneering work with schizophrenic patients. In the course of his research, Garmezy came across a curious finding: Even in the face of difficult circumstances, some of the adult schizophrenics he worked with had surprisingly functional lives. They held jobs, managed to keep their activities in order, and even maintained satisfying romantic relationships. These subjects—labeled "reactive" in Garmezy's study—stood in stark contrast to the "process" schizophrenics, who seemed to live life in a series of revolving doors between institutions, unemployment, and homelessness.

The differences between these two groups piqued Garmezy's interest, and he initiated a new experiment to study the children of schizophrenic parents. Much to his surprise, he found that 90 percent of the children exhibited normal functioning, including good peer relationships, academic achievement, and purposeful life goals. Garmezy urged his fellow clinicians to focus less on risk factors and more on "the forces that move such children to survival and adaptation." At his call, a fuller body of work on psychological resilience started to take shape in the early 1970s.

In the initial stages of this nascent field, social psychologists touted the strength and fortitude of their subjects. Phrases like "the invulnerable child," "superkids," and "vulnerable but invincible" were picked up by the media and news organizations, creating the impression that such children were endowed with extraordinary coping skills. The reality was more nuanced. As Ann Masten, Garmezy's protégé, described:

> The great surprise of resilience research is the ordinariness of the phenomena. Resilience appears to be a common phenomenon that results in most cases from the operation of basic human adaptational systems.

Masten went on to argue that if these basic human adaptational systems are protected and in good working order, most children will meet their developmental milestones, even in the face of great adversity.

Resilience was not the stuff of extraordinary superkids after all. It was commonplace. So much so, in fact, that Masten titled her paper "Ordinary Magic."

So why don't psychologists have a more widespread appreciation for our innate characteristic of resilience? The answer, it turns out, is rooted in the history of psychology and in our own cultural responses to adversity and trauma.

THE RESILIENT COHORT

In 1991, not long after finishing his PhD, newly minted clinical psychologist George Bonanno was offered a position to study bereavement at the University of California in San Francisco. He originally intended to spend a few years in the field of trauma and bereavement, familiarizing himself with the literature before moving on to what he imagined to be less morose territory. What he discovered, instead, was a body of research that included surprisingly little quantitative analysis.

"I kept arguing that we needed to be more precise," he said. "We needed to work with long-term prospective data in a way that tracked people empirically, using number crunching to calculate distributions of people. That was the only way to escape the confusion of older, more conceptual models."

Bonanno has spent the better part of two decades working with just these types of empirical models, studying how people cope with trauma and loss. The results have surprised everyone, most of all himself.

Throughout most of the twentieth century, the grieving process was filtered through the lens of Freudian psychoanalysis. Freud referred to grief as the "*work* of mourning," a term he introduced in a piece titled "Mourning and Melancholia," written at the height of the First World War. The "work of mourning" described the exhaustive process by which the libido detaches from the now nonexistent object of grief. In simplified terms, Freud theorized that if one didn't process each and every memory involving the deceased and then create some psychic distance from them, the mind would inevitably break down and one would begin to exhibit neurotic behavior.

As with so much of Freud's work, it's important to remember the context within which his ideas developed. His theories were rooted in intimate, individual observation, not statistical analyses of populations, and though they obviously reflected his own thinking, they also reflected a cultural backdrop of mass trauma and loss during the First World War.

These ideas exploring grief were relatively nascent in Freud's writings and they might have remained so if Freud's followers had not pulled together the threads of the theory. Psychoanalysts who treated patients suffering from grief expanded on Freud's ideas and, through their reinterpretations, the notion of "grief work" caught hold in Western culture. One of the most influential of these analysts was the American psychiatrist Erich Lindemann. In 1944 he published a seminal paper that proposed that all grief was delayed, festering in the unconscious and wreaking all sorts of havoc on the psyche. Lindemann believed that even if the bereaved seemed outwardly to be coping well with loss, his delayed grief would come back at some future time to haunt him.

Once these tenets of grief work took hold, various stage models started to appear, the most famous of which is Elisabeth Kübler-Ross's five stages of mourning: denial, anger, bargaining, depression, and an ultimate acceptance. While these stages are popularly understood to apply to all types of grief, in actuality, Kübler-Ross developed them based on her work with terminally ill cancer patients who were working to accept their own mortality, not that of another. She had never once tested the model with subjects grieving the loss of another person or experiencing other forms of trauma.

In sum, by the last decade of the twentieth century, the bulk of Western cultural assumptions about the grieving process were based on almost no quantitative research at all. There was essentially one conceptual model for processing grief, and any framework for human experience deviating from it was viewed as apostasy.

This was the world of ideas George Bonanno encountered when he accepted the position in San Francisco back in the early 1990s. Unlike his predecessors using only Freudian theory as a framework, Bonanno wanted to analyze how grief and trauma were expressed in populations, not just in individuals. Much akin to the way the marine biologists monitored the triggering events of coral reefs, Bonanno wanted to investigate people's psychological response to a short, sharp shock: What happens after a traumatic event pushes us from our everyday state of mind into a state of extremis?

To investigate these ideas, in 2002, Bonanno joined a project called Changing Lives of Older Couples (CLOC), initiated at the University of Michigan. The CLOC team conducted interviews with approximately 1,500 married people in the Detroit area over the course of a decade. Bonanno wanted to identify the participants who exhibited psychological resilience after the loss of a spouse.

This is not to say that he was looking for people who were completely unaffected by the death of a loved one or incapable of feeling sadness. Rather, Bonanno wanted to ascertain how many people actually experienced the delayed grief, denial, or theoretical stages of mourning that made up the bulk of our cultural understanding of bereavement.

To conduct his research, Bonanno needed to reach out to the CLOC subjects who lost a spouse throughout the course of the study. This narrowed the initial group of 1,500 down to 205 participants. After conducting a series of interviews with subjects after their loss—augmenting the information gathered from interviews conducted before the loss—Bonanno was able to the break responses down into five main patterns: (1) chronic depression, (2) chronic grief, (3) depressed-improved, (4) recovery from grief, and (5) resilient. As indicated by the pattern names, the chronic depression group suffered from pathology both before and after the loss. The chronic grief group functioned well preloss but were paralyzed by grief both immediately after the loss and several years later. The depressed-improved group experienced depression before the loss but reported a positive affect after the loss. The recovery from grief group experienced feelings of grief like yearning, shock, and anxiety that eventually subsided. And, last but not least, the resilient group experienced no significant trauma either immediately or several years after experiencing the loss.

The fact that these five patterns emerged did not surprise Bonanno. The real shocker was their relative distribution across the group. If the Freudians were right, then every one of those people suffering a loss could be expected, in the absence of grief work, to become dysfunctional. But only 25 percent of the CLOC subjects were debilitated with chronic grief or depression, and the number of subjects exhibiting delayed grief was so small—3.9 percent—that it registered as barely a blip in the statistical results. Of the remaining cohorts, 20 percent of the grieving participants recovered on their own while 45.9 percent reported no debilitating grief at all. This was the group, almost half of the population, that Bonanno labeled resilient.

It bears repeating that Bonanno was not defining resilience as a lack of feeling or absence of sadness. He used the term "resilient" to identify people capable of functioning with a sense of core purpose, meaning, and forward momentum in the face of trauma (echoing our own definition: "the capacity of a system, enterprise, or a person to maintain its core purpose and integrity in the face of dramatically changed circumstances.") By all accounts, the resilient cohort felt great sadness after their loss, and they were seriously challenged by the navigation of a major life change. But they described moving on—adapting and even growing from the loss—without experiencing the stages of grief or the consequences predicted by failing to do Freudian grief work.

In short, they bounced back.

These same patterns and sizes of cohorts have been identified in the wake of traumatic events like natural disasters and terrorist attacks. After the World Trade Center attacks on September 11, Bonanno and his research team did large-scale surveys with various groups of New Yorkers. The cohort that experienced the greatest difficulty, perhaps unsurprisingly, comprised those who had both witnessed the attacks firsthand and lost a loved one. Bonanno found that the incidence of posttraumatic stress disorder (PTSD) was higher in this cohort than in his randomly sampled groups—approximately 30 percent—but still didn't exceed a third of the group.

Over and over again—in natural disasters, after the SARS epidemic, following the loss of a child or spouse—Bonanno's longitudinal studies on loss and trauma revealed the exact same pattern at the population level. No matter how bad the trauma, rates of PTSD never exceeded one-third, and rates of resilience were always found in at least one-third and never more than two-thirds of the population.

"This pattern of response is so ubiquitous, and so consistent, it begs the question: Why are we, as a species, designed this way?" asks Bonanno.

One possible answer is that the design ensures that there is always at least a sizable minority, or even a majority, to take care of those deeply affected by a trauma.

Personal resilience has a dizzyingly long list of correlates. Although it would be impossible for us to give adequate treatment to all of them here, it is worth taking some time to go over a few of the more salient ones. Among these, innate personality traits like optimism and confidence have emerged as some of the most protective assets against life's stressors. Think of Bella and Jack at Lingfield orphanage: According to Moskovitz's study, they were able to charm the adults and function with self-agency at the orphanage, creating a positive feedback loop with the staff and their families that resulted in better and better care. This *ego-resiliency*—defined as the capacity to overcome, steer through, or bounce back from adversity—was first noted by developmental psychologists Jack and Jeanne Block in 1968, in a highly regarded longitudinal study documenting the lives of one hundred young adults over more than thirty years. In addition to ego-resiliency, the Block study measured a characteristic they called *ego-control*, or the degree to which an individual has the ability to delay gratification in service of future goals. Subjects exhibiting the combination of ego-resiliency and ego-control were better able to adapt flexibly to different circumstances and succeed in the midst of challenges.

Such personality traits are rooted in belief systems that allow one to cognitively reappraise situations and regulate emotions, turning life's proverbial lemons into lemonade. Social psychologists refer to this as *hardiness*, a system of thought based, broadly, on three main tenets: (1) the belief that one can find a meaningful purpose in life, (2) the belief that one can influence one's surroundings and the outcome of events, and (3) the belief that positive and negative experiences will lead to learning and growth. Considering this, it should come as no surprise that people of faith also report greater degrees of resilience.

Psychologist Kenneth Pargament has spent the lion's share of his academic career investigating the links between religion and resilience. In addition to offering all of the benefits of a community—including support groups and coping methods for people financially or socially disenfranchised—Pargament attributes the power of religion to its invocation of the sacred. His work specifically distinguishes between secular coping mechanisms and sacred ones, those that work in direct collaboration with a god by either creating a partnership or relying on the utter relinquishment of control. As anthropologist Clifford Geertz wrote in his seminal essay "Religion as a Cultural System":

> The strange opacity of certain empirical events, the dumb sense-lessness of intense or inexorable pain, and the enigmatic unac-countability of gross iniquity all raise the uncomfortable suspicion that perhaps the world, and hence man's life in the world, has no genuine order at all—no empirical regularity, no emotional form, no moral coherence. And the religious response to this suspicion is in each case the same: the formulation, by means of symbols, of an image of such a genuine order of the world which will ac-count for, and even celebrate, the perceived ambiguities, puzzles, and paradoxes in human experience. The effort is not to deny the undeniable—that there are unexplained events, that life hurts, or that rain falls upon the just—but to deny that there are inexplicable events, that life is unendurable, and that justice is a mirage.

This connection between religious faith (or, more broadly, a personal spiritual cosmology) and resilience presents an intriguing rejoinder to atheist critics of religious beliefs. While such beliefs may or may not be *true*, they may nonetheless be *adaptive*. That is, religious belief persists and thrives, in part, not because it necessarily guarantees persistence of one's soul in the next life, but precisely because it confers a measure of psychological resilience upon its possessors.

Of course, religious practitioners are not the only group that exhibits a high degree of resilience. Cultural identity also plays a role. For example, researchers found that Hispanics deemed at high risk by all the standard indicators appeared healthier as a group when they expressed a strong attachment to their Hispanic heritage. Such a finding suggests that members of a culture affirming strong in-group loyalties will exhibit greater personal resilience, a point we will revisit in greater depth in our coming discussion of cooperation.

The ability of members from certain communities to bounce back from adversity is also aided by high-functioning social networks— friends, family, religious and community organizations, satisfying jobs, and access to government support and resources. A seminal forty-year longitudinal study by researchers Emmy E. Werner and Ruth S. Smith, published in 2001, followed nearly seven hundred children growing up in Hawaii with risk factors like poverty, parental discord, and prenatal stress. Werner and Smith concluded that social factors—such as the support of an adult role model in the community—buffered the effect of adversity and appeared to predict positive outcomes in anywhere from 50 to 80 percent of their high-risk population. In 2000, researchers at the University of Maryland, College Park, showed that access to social resources like a supportive relationship with a teacher and a variety of well-organized extracurricular activities correlated with high academic achievement. Conversely, their studies suggested that children's exposure to violence—primarily intrafamily and neighborhood—had significant negative effects on their mathematics and reading performance on a standardized exam.

Social resources are the oil that greases the wheels of well-functioning social networks. And a flurry of new research suggests that these networks can even have a physiological impact. Psychologists Sarah Pressman and Sheldon Cohen at Carnegie Mellon found that college freshmen with larger social networks had a stronger immunological response to getting a flu shot, while Alexis Stranahan, David Khalil, and Elizabeth Gould from Princeton University found the converse: that social isolation can reduce the physical benefits of exercise. Socially isolated rats sprouted fewer new neurons and neural connections as a result of wheel running than rats living in groups. Social isolation is not just bad for our psychological well-being. It appears to leave its trace at the cellular level.

All of these, and many other, correlates to personal resilience are rooted in our beliefs and our experiences. Whether cultivated through wise mentors, vigorous exercise, access to green space, or a particularly rich relationship with faith, the habits of personal resilience are habits of mind—making them habits we can cultivate and change when armed with the right resources.

This brings us to another, less appreciated aspect of personal resilience: the influence of genetics. The sequencing of the human genome in the last decade has led to a flurry of speculation that scientists would one day identify genetic triggers for undesirable traits like a depression gene or an alcoholism gene. Today our understanding of behavioral genetics is more nuanced. We recognize that the emergence of traits and behaviors is informed by the dynamic interplay between genes, lived experience, and triggers in the environment. Behavioral geneticists call this framework *genes–environment interaction* (GxE).

About thirty-five years ago, researchers from the University of Otago in Dunedin, New Zealand, initiated a longitudinal study with more than one thousand infants across New Zealand. This birth cohort was then assessed every two years or so on many different life factors, creating a data set that, like Bonanno's CLOC study, was a gold mine for behavioral scientists doing population studies. In 2003, researcher Terrie E. Moffitt and her husband and co-investigator, Avshalom Caspi, used the Dunedin study to investigate the impact of a gene called *5-HTT,* which helps regulate the transmission of serotonin, a neurotransmitter implicated in a wide variety of mood disorders. (Antidepressants like Prozac and Zoloft, for example, target serotonin and its ability to transmit nerve signals in the brain.)

The *5-HTT* gene has two variations, called alleles, and each allele occurs in either a long version or a short version in a wide variety of species, including human beings. In previous studies with mice and monkeys, animals with two long *5-HTT* alleles coped better with stressful situations than those with two short alleles or one short and one long. The mice with two copies of the short allele were much more fearful when encountering loud noises, and the monkeys with short alleles had impaired serotonin transmission when raised in a stressful environment.

Moffitt and Caspi were curious about how this gene might be distributed in the New Zealand birth cohort and analyzed 847 members for stressful events like a death in the family, job loss, and romantic breakup. Out of those subjects who met the criteria for recent stressful events, a whopping 43 percent with two short alleles reported experiencing depression versus only 17 percent for subjects with at least one long allele. Caspi and Moffitt concluded that the shorter variant of *5-HTT* rendered subjects vulnerable to adversity, while the long allele acted as a buffer. Without the environmental trigger, however, they speculated that the difference between short and long alleles mattered little in the subjects' lives.

Geneticists call this theory *stress diathesis* or *genetic vulnerability*— the idea that certain gene variants can increase a person's propensity toward depression, anxiety, and any number of other pathologies like antisocial and sociopath behavior, if—and scientists make a point of emphasizing this *if*—the subject encounters potentially traumatic or stressful life events. Caspi and Moffitt's study emphatically did not prove the existence of a depression gene. Rather, the study used empirical analysis to show how GxE might create depression vulnerability.

Subsequent meta-analysis (i.e., studies of the studies) of the correlation between personality traits like neuroticism and depression and variations in the *5-HTT* gene have been mixed (due perhaps to differing methodologies of the underlying studies involved), suggesting that the correlation may be smaller than originally found. Yet research continues to find some persistent correlation between *5-HTT* and specific traits like optimism and happiness, which suggests that the long and short allele variants may be among a host of many hidden factors that cause the Berls and Leahs of the world to have a different life experience than the Jacks and Bellas.

So what can we do for them? Genetic correlates to resilience are illuminating from a scientific perspective but they don't offer tangible benefits to the cohort suffering from trauma in Bonanno's studies. After all, we can't control our own genes. And even if we could, it's impossible to say what triggers might set off our as-yet-untested vulnerabilities.

The standard tool kit for treating trauma includes pharmacological intervention and intensive therapy. These are important tools, often essential to the healing process, and we heartily endorse their appropriate use. But they may not be accessible—financially or logistically—to everyone. They also may not be the right intervention for every circumstance, and, we hope it's uncontroversial to say, it would be better if other interventions could obviate their need in the first place.

Current research in neuroscience is revealing the effectiveness of one tool in particular that can complement other forms of intervention: This tool is portable, teachable, free, and it's been on the market for more than two thousand years. It's called mindfulness meditation.

EMOTIONS AND RESILIENCE

Before we look at how we might use this tool, we need to slightly adjust the way we conceive our own emotions. For most of us, emotions are things that happen to us. We might be going happily about our day when, in the presence of some small social injustice or irritant, we're hit by a wall of anger. Boom! We feel upset. The relationship between the event and our emotion is one of causation: The event causes us to feel a certain way. It's not like we have any choice in the matter. Other times, our emotions creep up on us, but no less involuntarily. We might wake up in the morning feeling bright and cheerful until, slowly, through-out the day, a veil of sadness descends upon us. These things—our emotions—seem to function of their own accord. When they strike, we feel little control over them.

Researchers who study mindfulness and attention often conceive of our emotions differently. In their view, emotions are not things that happen to us. Rather, they exist—metaphorically, of course—as a kind of psychic currency, held in reserve. When we waste this reserve—giving over our attention to every single distraction from the outside environment—it dwindles down into an empty account, and we are left feeling fatigued or, worse, in a downward spiral of negative affect like anger or despondency. With practice, on the other hand, we can train ourselves to spend deliberately and judiciously, keeping us from drain-ing our own emotional coffers.

Mindfulness meditation training is the tool that lets us do so. It allows us to take more intentional control over our emotions. Some of this training is drawn from Eastern religions—specifically Buddhism—but we present our findings with a completely secular intent: These tools of meditation, mindfulness, and increasing awareness have been proven to aid in resilience training. And, unlike other tools that depend on innate characteristics, genetics, and social resources, these training exercises are entirely under the control of the individual using them.

There are many systems of meditation, so the following examples are by no means exhaustive, but they provide a framework for discuss-ing the different ways in which our brains regulate attention. Medita-tion experts often refer to two different styles: focused-attention and open monitoring. Focused-attention meditation maintains attention on a specific object of concentration; when thoughts and sensations arise, the mind allows them to pass without clinging to them and then brings itself back to focus on the chosen object. This process—what we will refer to later as "detachment"—cultivates the presence of an internal witness observer, capable of stepping back and disassociating from the environment, focusing instead on the chosen object.

In open monitoring, on the other hand, the object of focus recedes and a sustained awareness of all sensory experience is cultivated. Open monitoring—what we will later refer to as "attending"—is characterized by an open, present, and nonjudgmental awareness of stimuli in the environment.

There is a third type of meditation that will play the pivotal role in our story of personal resilience. It is often referred to as "loving-kindness," or a practice of compassionate meditation. This is the technique of cultivating greater empathy through meditation, beginning first by focusing on loved ones and then expanding the focus of compassion toward all beings. Such practices, performed by meditation masters, produce significant activity in the insula—a region near the frontal portion of the brain that plays a key role in bodily representations of emotion—as well as the temporoparietal junction, an important part of the brain for processing empathy. One man in particular has been fundamental to bringing more attention to this, as well as other practices, casting his scientific lens on the contemplative arts so that researchers might appreciate meditation's ability to radically alter our brain chemistry.

Neuroscientist Richard Davidson is tall and lanky. If you squint while looking at him, you might even mistake him for a graduate student with that slightly disheveled hair and suede-elbowed jacket. In fact, Davidson is sixty years old, and he is both a distinguished scientist and an avid practitioner himself of mindfulness meditation. When Davidson was at Harvard in the early 1970s, he became interested in incorporating these and other contemplative studies into his graduate work in psychology and neuroscience.

"When I approached my advisers with the idea," Davidson said, "they told me, in so many words, that I would never have any kind of career with that focus. They felt that I would never be taken seriously as a scientist."

Instead of tossing aside his research interests, Davidson took a life-changing journey to India in 1974. He immersed himself in contemplative traditions through a meditation retreat and then returned to Harvard to complete a doctorate in biological psychology. He has been at the University of Wisconsin in Madison ever since, and today he directs both the Waisman Laboratory for Brain Imaging and Behavior and the Center for Investigating Healthy Minds.

In many ways, Davidson's journey as a researcher mirrors that of George Bonanno. Before the 1990s, his field of study—psychology—was primarily interested in negative emotions. The research was slanted toward emotions like fear, anxiety, and disgust. But Davidson was interested in looking at positive emotions like kindness and compassion.

"Twenty or thirty years ago, compassion wasn't on the scientific map whatsoever. And yet, compassion is in our repertoire of behavior and in the repertoire of nonhuman primates as well. It appears to be something fundamental to who we are. To not have a serious inquiry into that fact was very limiting."

In 1992, the Dalai Lama invited Davidson to India to meet with monks with extensive experience in the contemplative arts. This was the beginning of a long and fruitful collaboration between Davidson's lab and a rotating cast of Tibetan Buddhist monks—the gold-medal champions of meditation practice—who regularly visit him in Madison and participate in his experiments.

In June 2002, Matthieu Ricard, a monk from the Shechen Monastery near Kathmandu, entered Davidson's lab. Having spent tens of thousands of hours in meditation practice, Ricard is a mindfulness expert. But he is also a scientist, a graduate of the Institut Pasteur with a PhD in molecular genetics, as well as the son of famous French philosopher Jean-François Revel. His unusually diverse background gives him the tools to bridge conversations between the scientists in Davidson's lab and practitioners of the contemplative arts like his own community of monks.

Ricard was instructed to meditate on unconditional loving-kindness and compassion while his brain waves were analyzed with fMRI, functional magnetic resonance imaging, the first experiment to ever use such equipment in service of analyzing meditation's effect on the brain. What followed continues to awe Davidson and his fellow researchers. Ricard's brain waves started oscillating at approximately 40 cycles per second, indicating powerful gamma activity. Gamma waves—underlying the highest of our mental activities, including consciousness—are usually difficult to detect, but Ricard's were prominent. Even more remarkable, the EEG output also showed a synchronization of the oscillations throughout Ricard's cortex. As this is a common phenomenon in patients under anesthesia, it suggests that Ricard is able to block out sensations of pain through thought manipulation alone.

Lest we forget, Ricard's brain was not commonplace—he has been rigorously training it over the course of several decades. Like body builders who decide they need to work on their biceps, Ricard and his fellow monks have spent most of their lives intentionally changing their conscious thoughts through contemplative practice. The results in Davidson's lab showed unequivocal changes in the physiology and behavior of the monks' brains. Monks with more than ten thousand hours of practice showed significantly greater activation of their limbic systems, indicating that their brains had changed to reflect their focus on empathy. Lab results also showed that their baseline brain functioning was altered even when outside of a meditation practice, suggesting that the brain changes were permanent.

For Davidson, the research was empirical confirmation of something he had been trying to investigate for years: the transformational powers of meditation on the brain. Until recently, the thrust of most psychological research on temperament held that once an individual reached the post-adolescent, young adult period, he or she was relatively static. Whether the person was categorized as fearful, angry, or shy, psychologists theorized that these characteristics would be constants. But the EEG scans coming out of Davidson's lab served as a corrective to arguments for fixed traits and attributes.

"Our brains do change," Davidson explained. "This is a concept we now know in neuroscience as neuroplasticity. But the possibility of transformation, even radical transformation, has always been implicit in the Buddhist traditions."

Over the last twenty years, countless studies have emerged illustrating the ways our minds change in response to activity and life experiences. Brain scans have shown that taxi drivers in London have a larger-than-average hippocampus—the part of the brain associated with navigation—compared with bus drivers, perhaps because they have to hold so many routes in their minds. And professional musicians have been found to have higher volumes of cortical matter—areas of the brain that engage while playing an instrument, including motor regions, anterior superior parietal areas, and inferior temporal areas—compared to nonmusicians.

But what about the rest of us? If meditating monks—experts at mindfulness—can train their brains to focus on positive emotions like compassion after years of practice, is it possible for novice practitioners to gain the benefits as well? Can meditation serve as a resilience booster, inoculating the brain against anxiety and stress and, possibly, genetic influences like the variants of the *5-HTT* gene?

In January 2011, researchers at the Massachusetts General Hospital, led by Dr. Sara Lazar, reported a suggestive finding. Lazar took magnetic resonance imaging (MRI) scans of the brains of novice subjects, who then participated in an eight-week meditation program. At the end of the eight weeks, after using a form of open monitoring meditation for an average of twenty-seven minutes a day, Lazar and her team scanned the subjects again. They found measurable changes in the regions of the brain associated with self-awareness, compassion, and introspection, including increased gray-matter density in the hippocampus, known to be important for learning and memory, as well as decreased gray-matter density in the amygdala, which is known to play an important role in anxiety and stress. Control subjects who did not receive the training had no meaningful changes in their brain physiology.

"These data provide the first structural evidence for experience-dependent brain plasticity associated with meditation practice," reported Lazar. And, unlike the meditation experts studied in Davidson's lab, these subjects were entirely new to the practice of meditation, suggesting that the brain can change relatively quickly.

Lazar's research on meditation and neuroplasticity is mirrored by provocative research that suggests a complementary link between meditation, well-being, and physical longevity at the cellular level.

Each cell in your body except your sex cell has twenty-three pairs of chromosomes; one set each from your mother and father. These chromosomes package your genes—the long-form instructions, written in the language of DNA, for making you. A cooking metaphor is incomplete but instructive: You can think of each base pair of DNA molecules as a single line in a recipe for a specific dish; each gene is equivalent to the completed recipe; each chromosome is akin to a cookbook of such recipes; and all the chromosomes together like a complete library of French cooking.

At the end of these chromosomes are telomeres—stubs of DNA that, like the protective endpapers at the joints of a cookbook, keep the spools of DNA they contain from unraveling. Each time most of the different types of cells in our bodies replicate in order to make new copies of themselves, the telomeres guarding their chromosomes lose a bit of their DNA. After many replications, the telomeres are worn away entirely; soon after, the cell dies.

In 2004, University of California, San Francisco, brain researcher Elissa Epel and her colleagues found that chronic stress also chips away at telomeres, reducing the maximum number of times that a cell can reproduce and accelerating the aging process.

Epel's team studied women aged twenty to fifty who had experienced the lasting stress of taking care of a child with a severe chronic illness, such as autism or cerebral palsy. They also studied a control group of women whose children were healthy. The research team found that the longer a woman had been caring for an ill child, the shorter her telomeres and the lower her level of telomerase, the enzyme that maintains the length of telomeres. In addition, the greater each subject's perception of her stress, the worse she scored, across the board. Women with the highest perceived stress had the telomeres of a woman a full decade older. (This might be why so many U.S. presidents, for example, leave the office looking like they've taken an accelerated aging pill.)

In 2010, a group led by UC Davis researchers Tonya Jacobs and Clifford Saron, and which included Elissa Epel, concluded a study that suggests a possible reverse correlation—between meditation, mind-set, and enhanced longevity. In the study, a group of thirty subjects were each given six hours of meditation a day for three months. During the training, subjects were instructed in techniques of focused attention—attending to the mind's activities in a nonjudgmental way—and the generation of benevolent mental states, such as compassion, empathy, and equanimity.

After the meditation training, the subjects were compared to a control group of subjects matched for age, sex, body mass index, and prior meditation experience who had been waitlisted for the same meditation training (and then subsequently received it).

The subjects who had meditated were found to have a dramatically increased sense of mindfulness (being able to observe one's experience in a nonreactive manner), purpose in life (viewing one's life as meaningful, worthwhile, and aligned with long-term goals and values), perceived control (over one's life and surroundings), and decreased neuroticism (negative emotionality) when compared to the control group.

All of this might be the expected, natural psychological consequence of meditation. However, when the retreat group's blood was analyzed, it was also found to have substantially increased telomerase—that enzyme responsible for maintaining telomeres. Telomerase activity was about one-third higher in the white blood cells of participants who had completed the retreat than in the matched group of controls.

This was particularly strongly correlated among subjects who reported that, during the retreat, they found an increased sense of their purpose in life. "The take-home message from this work is not that meditation directly increases telomerase activity and therefore a person's health and longevity," says Saron. "Rather, meditation may improve a person's psychological well-being, and in turn these changes are related to telomerase activity in immune cells, which in turn has the potential to promote longevity in those cells.

"In other words, activities that increase a person's sense of well-being may have a profound effect on the most fundamental aspects of their physiology," he adds. "It doesn't necessarily have to be meditation per se; it's really about creating conditions in which you can flourish and your purpose can come into being."

It's important to note that these findings are suggestive—not definitive. Jacobs, Saron, and their research team undertook a snapshot in time, not a longitudinal study. But the results are tantalizing.

While Davidson's lab continues to study the cultivation of compassion and kindness through meditation, Lazar's lab finds evidence for its physical impacts on the brain, and Jacobs, Epel, and Saron uncover links between the meditating brain and the mechanisms of cellular aging, research scientist Raffael Kalisch at the University of Hamburg is interested in using detachment techniques as a potential tool for pain management. In one of his experiments, he attached subjects to wires before telling them that they would receive a painful electric shock at some point in the next fifteen seconds. Normally this kind of announcement would instigate physiological and psychological responses: increased heart rate, perspiration, maybe difficulty breathing, perhaps even a sensation of panic. Imagine that moment at the doctor's office, right before you get a shot: *Now this won't hurt a bit . . .*"

Kalisch and his team, however, trained their experimental subjects in the practice of detachment, based on mindfulness meditation exercises.

"We encouraged our subjects to imagine themselves in a geographically distant position, to look at the situation as if they were standing outside of it," Kalisch said. "They were instructed to tell themselves that the stimuli did not concern their inner core self."

The detachment thought process—involving cognitive reframing of the situation—goes against most of our primal instincts. In the face of an imminent attack on our body—pain, shock, violence—most of us feel threatened. Kalisch trained his subjects to reframe the painful shock as irrelevant and to conceive of their core selves as unaffected.

The intervention worked: His team showed that the detachment techniques reduced anxiety (reflected by physiological measures like heart rate and skin conductance) when compared with control subjects immersed in the experiment. Such results suggest that we can utilize withdrawal and detachment exercises when we feel threatened in daily life or overwhelmed by stress. Much like Matthieu Ricard's brain was able to achieve a state similar to an anesthetized subject's, these kinds of detachment techniques can form the basis of long-term pain management strategies. Once these habits of mind are firmly established, they are that much easier to maintain due to their well-trodden neural pathways in our brains.

Such findings offer promise for those looking for a long-term strategy for pain and anxiety management, but certain jobs and situations are so extreme that detachment is not a viable option. Imagine you're a firefighter or an ER doctor: You've been working for eighteen hours without a break and you may not have anything left in your emotion account to reframe the situation. There is no time to try to find a mental health worker—much less sit down for a session with one—and many of the psychopharmaceuticals on offer have side effects that might impede your cognitive functioning. These are the moments when a meditation practice like attending confers an advantage, bringing support to places that other interventions simply can't reach. Amishi Jha, professor of psychology at the University of Miami, is teaching these techniques to workers who struggle with on-the-job stress every single day: United States military personnel.

Jha's project, entitled the STRONG Project (Schofield Barracks Training and Research on Neurobehavioral Growth), is working with a small group of Marines both before and after deployment. She is looking for an alternative to mindfulness training techniques like detachment that exhaust both the emotional coffers and executive functioning skills like decision making and spatial recognition, what she refers to as the "working memory capacity." In a combat soldier, Jha argues, this capacity becomes impaired through stress, lack of sleep, physical duress, and anxiety. Her training techniques help exhausted soldiers stay mindful and present without asking them to reappraise their experience or detach from their current situation.

"All of the cognitive reframing required for other meditation practices involves tapping into an already drained tank, further depleting them," Jha said. "It might actually make things worse."

Jha's research suggests that the practice of attending will better help soldiers maintain emotional control in the midst of a cognitively demanding set of circumstances like warfare. And, although the sight of men in army fatigues sitting in the lotus position and staring at their nostrils does not call to mind a warrior's pose, Jha maintains that it is the very foundation for every soldier's effectiveness on the battlefield.

"I keep telling the generals, look: the things you want these guys to do—use their weapons appropriately, find the right route, etc.—are the exact same things that will require them to regulate their emotions. They can't do these things if they're freaking out and shooting each other. This kind of training will, we hope, bring together the two sides of the same domain: emotions on the one hand and the hard-core mission skills on the other."

Whether choosing long-term meditation practice or attention-focusing techniques in moments of high stress, mindfulness serves to bolster an individual's psychological resilience with a tool that's portable, teachable, and free. Most encouraging of all, it has been proven to work, over and over again, not only with the relatively healthy cohort that Bonanno identified throughout his studies—the Jacks and Bellas of the world—but with those susceptible to or already suffering from trauma. For the Berls and Leahs in our communities, individual resilience can become a habit of the mind.

Yet even the hardiest individual cannot go it alone—our resilience is rooted in that of the groups and communities in which we live and work. In turn, at the core of social resilience are two factors that we'll now explore in some detail: cooperation and trust.

5

COOPERATION
WHEN IT COUNTS

In 1906, British physiologist and pharmacologist Sir Henry Dale made a most unusual discovery. He had recently started experimenting with the liquefied extract of the human posterior lobe of the pituitary gland, the kidney-bean-sized organ at the base of the brain. Dale suspected that the gland's hormonal secretions played an important role in the onset of labor in pregnancy, so on a hunch he injected the extract into the uterus of a pregnant cat.

Dale's hypothesis proved true and then some: His feline subject delivered her litter of kittens in record time. Appropriately, he named the drug using a combination of Greek words: "fast" (*oxy*) and "birth" (*tocos*). Sir Henry Dale had discovered the hormone oxytocin.

When word got out about Dale's discovery, a whole industry of labor-inducing drugs was born. Slaughterhouses began to harvest the postpituitary extracts of cattle and sell them to physicians under the name Pituitrin. When adverse reactions were reported from Pituitrin, pharmaceutical companies took over and started developing chemically purer formulas. By 1953, chemist Vincent du Vigneaud, riding a wave of acclaim after successfully synthesizing penicillin, used his knowledge to fully isolate and synthesize Dale's "fast birth" discovery: The chemistry of birth would now be available in IV drip form. By the 1960s, physicians were administering oxytocin to millions of pregnant women under a pharmaceutical name familiar to anyone who sets foot in a hospital delivery room today—Pitocin.

For many decades, it seemed that oxytocin's main claim to fame would be its role in bringing babies into the world. More recent findings, however, reveal this hormone has a much bigger and more nuanced role to play in human nature. Oxytocin is among the most important neurochemicals involved in our experience of two linked human traits that are essential for our survival: trust and cooperation.

As we will see throughout this chapter, resilience is predicated on trust in a system, allowing potential adversaries to move seamlessly into cooperative mode—and quickly, during the moments when it counts the most.

33 LIBERTY STREET

On Friday, September 12, 2008, the treasury secretary of the United States, Henry "Hank" Paulson, and a team of officials from the Federal Reserve called thirty or so of Wall Street's most powerful executives to the Fed's offices in Lower Manhattan for a secret evening meeting. Limousines, town cars, and taxis pulled up to the giant slab of white stone at 33 Liberty Street just before six o'clock. Some of the lower-level bankers got out in front of the building, making their way through the onslaught of press. Savvier senior executives opted for the building's underground garage, to avoid the scrum of reporters.

There was good reason for their low profiles: All at once, and with a frightening speed and apparent randomness, seemingly unassailable emblems of the U.S. financial industry were being felled like giant sequoia in the forest. Just one week earlier, Washington had seized control of the failing mortgage giants Fannie Mae and Freddie Mac. Months prior to that, government officials had brokered a deal to save Bear Stearns, agreeing to put up $30 billion to help complete an acquisition of the failing bank. Bankers in the business for twenty and thirty years described the sequence of events leading up to the September 12 weekend as "scary," "terrifying," and "extraordinary." Alan Greenspan, the former chairman of the U.S. Federal Reserve, called the financial crisis the worst of his career "by far." Suddenly, even monolithic institutions like Merrill Lynch—the populist brokerage outfit with branches in almost every city in America—seemed dangerously close to collapse.

None more so than Lehman Brothers, the storied firm that had suffered a slow but steady decline over the course of the past year. As we discussed previously, Lehman was heavily exposed to troubled real estate investments—caught like so many others in the subprime mortgage debacle described in chapter 1, and though the firm's leaders had tried to raise fresh capital, they had made little headway. By Monday, September 8, Richard Fuld, Lehman's CEO, received the proverbial nail in the coffin: A possible rescue plan that Lehman had been cobbling together with a Korean bank had collapsed. When the market caught wind of the news, Lehman's shares plummeted 41 percent—this after falling 90 percent from a peak at the beginning of 2008.

By Friday, September 12, Paulson and his team—including then head of the New York Federal Reserve Tim Geithner and Federal Reserve chairman Ben Bernanke—were trying to navigate the waves of panic roiling the marketplace. Despite their previous bailouts over the course of the year, the federal government wanted all the bankers to understand that they were indeed willing to let Lehman Brothers fail.

At 6:00 p.m., the group of twenty or so financial titans—including John Mack, chief executive of Morgan Stanley; John Thain, head of Merrill Lynch; Vikram Pandit, chief executive of Citigroup; Lloyd Blankfein, of Goldman; and Jamie Dimon, head of JPMorgan Chase— were all settled in and seated around the table. Paulson declared his position to the players: It was time to reestablish moral hazard. The bankers would have to clean up their own mess. With no political will for another bailout, the only way to save Lehman Brothers—and the stability of the market, if not global capitalism itself—would be through the collective effort of the bankers sitting around the table. "You have a responsibility to the marketplace," Paulson told them.

A show of such large-scale cooperation was not unheard of. Most of the bankers around the table remembered an eerily similar message delivered ten years earlier by Geithner's predecessor, William McDonough. In 1998, the lauded hedge fund Long-Term Capital Management had collapsed, threatening to send cascading defaults throughout the markets. At McDonough's urging, a group of financial titans, including some of the same names and faces that met at 33 Liberty Street, ultimately agreed to join forces to bail out the firm, saving the entire financial system from destructive turbulence in the markets. Over the September 12 weekend, however, the answer would not be nearly so straightforward.

To begin with, no one was entirely certain of the government's resolve. Although Paulson and Geithner laid out their case adamantly, there was still some speculation that the Friday night session might really be a big game of poker. Were they bluffing? Surely, if push came to shove, Paulson and his team would blink and the threat of market discipline would give way to a quick infusion of cash. After all, they had once offered up the same claims of austerity for Bear Stearns before ultimately caving in. When all the posturing and grandstanding over moral hazard was over, wouldn't Paulson realize that Lehman Brothers was simply too big to fail?

Not only was there a sense of distrust in the government's position, there was also a sense of distrust in the current state of the market system. CEOs like John Mack of Morgan Stanley and Lloyd Blankfein of Goldman Sachs, who, only a month earlier, were considered immune to the toxicity of the subprime mortgage debacle, had found themselves vulnerable. The global financial markets were beginning to look like a line of precariously placed robust-yet-fragile dominos, one default pushing down another in a chain of interlinked interests and overleveraged debt packages, failures followed by bad faith followed by more failures. On Friday, September 12, the bankers did not fully understand how Lehman Brothers' toxic debt fit into the greater financial banking ecosystem. Even more frightening, no derivative formula or whiz-kid risk assessment manager could tell them how their own banks were doing in the midst of such dramatic and turbulent change. The players sitting around the table at 33 Liberty Street didn't just distrust the word of the government; they were having a proverbial crisis of faith in the whole system of global capitalism. Were the gods of the market dead?

By 7:00 p.m. on Friday, the government officials were laying out a few potential courses of action for the bank CEOs. One involved an orderly dismantling of Lehman that would allow the firm to die through bankruptcy. For purposes of simplicity, we'll call this option Let Lehman Fail.

Paulson and the regulators then offered an alternative: What if the Wall Street firms joined forces to take on Lehman's most toxic assets? The failing bank was a bit like a sports car that sat still in the driveway: It impressed from a distance but one peek under the hood revealed a lemon of a motor. What if they all worked together to make Lehman's motor look as good as the rest of its body? This option would call for the banks to form a team, pool their resources, and, ultimately, sweeten the deal for an eventual Lehman buyer. We'll call this option Teamwork Saves the Day.

This latter option would not come cheap. Though the costs would be shared among the players, the effort would require underwriting $85 billion of Lehman's most toxic assets—a pretty penny even in the best of times, never mind amidst the gathering financial crisis. More important, it would demand deep cooperation from the players—executives who, under normal circumstances, were fierce rivals. Could the bankers in the room at 33 Liberty muster the strategic presence of mind to join forces to save Lehman Brothers—and themselves? In a crisis of faith, would they be able to make a choice that served not their own best interests but the interests of the whole system?

The government emphasized the urgency of the situation. Virtually all of the institutions in the room were connected either directly or indirectly to Lehman's toxic assets; if Lehman went down, the hairball of its interconnected liabilities would ensnare the market as a whole. Timing was also of the essence: A decision needed to be made in a matter of days. As the sun started setting, Paulson's gravelly voice hung in the air: "Everybody is exposed."

But lingering doubts and questions remained: Both Bank of America and Barclays had expressed interest in acquiring Lehman Brothers at a fire sale price, and both of them had representatives present in the room at 33 Liberty Street. That left a bad taste in the mouths of the other CEOs. First, they were being asked to pony up a lot of cash for someone else's lemon of a motor and then, to add insult to injury, when the whole crisis was over, they would be handing over a fixed-up sports car to Bank of America or, even worse, Barclays: a direct competitor that was not even American.

By 8:00 p.m., the initial meeting was coming to a close. Paulson and his team had presented the options. The Wall Street executives had asked their questions. And now all of them had the night to think over their respective positions.

At 9:00 a.m. on Saturday morning, a most dangerous game of "Will We Save Lehman Brothers?" would officially begin.

It's worth taking a step back to look at the structure of the meeting at 33 Liberty Street described thus far. Structures create their own behavior, so it's no surprise that a meeting called in secret, convened under unclear rules, and rooted in a lack of transparency about each one of the individual banks' books would breed suspicion and distrust.

What's more, with the prospect of looming carnage, there was still the lingering possibility that one or more of the banks could really clean up on the deal. In the midst of losers, the bankers thought, there might also be some winners. But only if they played their cards right.

With such a start to the weekend, Teamwork Saves the Day was the underdog option. And yet the urgency and magnitude of the crisis had everyone's attention. If ever there was a day for inspired leadership to guide the emotions in the room, Saturday September 13 was it.

On Saturday morning, the CEOs and their closest advisers were ferried back to 33 Liberty Street in their town cars. Once again upstairs in the meeting room, the bankers broke up into groups to discuss the options placed before them, Let Lehman Fail and Teamwork Saves the Day. One Teamwork scenario group discussed the possibility of having every bank borrow from the Federal Reserve under the emergency lending provision it had recently started to offer. With their newly borrowed cash, the banks would be able to buy up Lehman's toxic debt, preventing it from bankruptcy and protecting themselves from exposure to the company's failing assets.

Another Teamwork group focused on helping Barclays or Bank of America buy Lehman's good assets, placing the toxic debts in a bad bank stuffed to the brim with $85 billion in investments gone wrong. Once that had been accomplished, the Teamwork group proposed that all the other Wall Street firms would inject sufficient capital into the bad bank to keep it afloat and unwind the debts over time.

With two possible cooperative solutions on the table and the memory of the Long-Term Capital Management lingering in everyone's minds, there were several possibilities for Teamwork to come together and save Lehman. Yet with every step forward toward cooperation, the level of distrust mounted. Would Paulson cave in? Would they even have the liquidity in their own coffers to hand over money to another firm? And would the system reward them with stability if they took such a chance? Perhaps most important, would this be the end of the crisis? Would the government secure them against possible failures in the future?

As the day progressed, grumbles of resentment started to drown out the expressions of fear and confusion. Several CEOs openly questioned why they should bear the cost of Lehman's problems when others who also faced exposure—institutional investors, hedge funds, and foreign investors—were not being asked to do the same. Even though they hadn't been the only proverbial binge drinkers at the bar, the bankers felt they had been left with the tab for the whole industry's decade-long bender. And if they paid up this time, wouldn't they just get suckered into paying next time too? Morgan Stanley's CEO, John Mack, expressed this suspicion to the room: "If we're going to do *this* deal, where does it end?"

With little faith that the government would protect them and even less faith that the system would reward them for choosing the collective good over their individual firms, the bankers were left in a stalemate on Saturday afternoon. In a market system in which big bets were typically backed by reams of data, it was the soft value of trust that would win—or lose—the day. But trust cannot just be turned on and off like a light switch.

Or can it?

Paul Zak, founding director of the Center for Neuroeconomics Studies at Claremont Graduate University, is one of several scientists starting to investigate the neurobiology of trust. His most recent research offers proof that our brains actually do have a finely honed on/off switch for the soft value of trust. The electricity powering this switch is none other than oxytocin—the neurotransmitter first discovered by Sir Henry Dale some hundred years earlier. Zak and his research colleagues demonstrated the role oxytocin plays in negotiations with an oft-used economics experiment called the Trust Game.

The game always involves two people, the subject and the stranger—someone the subject has never met. It also involves the diva of all lust and trust interactions: money.

In a control version of the game, the subject of the experiment is asked to withdraw money from a computerized bank account and then give some of it to the stranger (without any face-to-face contact). The subject has been informed that on receipt of the funds, the stranger will reciprocate, returning either the same amount or an even greater amount of money, at a later date.

In Zak's version of the experiment, however, a research team (headed up by Zak and Ernst Fehr of the University of Zurich) took 194 male students and instructed them to inhale a dose of oxytocin formulated as a nasal spray. The researchers compared their behavior with subjects who inhaled a placebo. Both the control and the experiment groups were then given ten dollars and told that they could share some, none, or all of it with a second person. If they did offer any portion to the second person, the amount dispersed would be tripled. The second person would then be asked to give as much back to the original subject as he felt inclined to give.

If the subjects trusted their recipient to be fair and give back half of the final total, both would gain. Without trust, however, their more obvious choice would be to hold on tight to the ten dollars and not even attempt to engage in the potential deal. The experiment shared some striking similarities to the bankers' predicament: If they all put money in to save Lehman, they would have a shot at collective success: market stability. But they needed to trust that they would be rewarded for this choice, by both the system and the government.

The experiment results were striking: Subjects who received the oxytocin nasal spray gave about 17 percent more money to the strangers than did the placebo-sprayed control group. Even more remarkably, twice as many oxytocin-treated subjects exhibited maximal trust by choosing to give their *entire* cash amount to the second person.

The team concluded that "oxytocin specifically affects an individual's willingness to accept social risks arising through interpersonal interactions." Their experiment became so well known that it led *New Scientist* journalist A. C. Grayling to offer a suggestion: "Instead of pumping billions into the money markets, governments might more cheaply have sprayed oxytocin up the noses of bankers and speculators."

•   •   •

By 5:00 p.m. on Saturday, the group at 33 Liberty Street was at an impasse. The Bank of America representative made it clear to Paulson and his team that they were pulling out of the running to buy Lehman Brothers, choosing instead to work out a deal with John Thain's Merrill Lynch. This left only Barclays as a possible buyer, but they refused to move forward unless the other players agreed to buy up Lehman's toxic debt. Mistrust was breeding mistrust and emotions in the room were running high.

Not surprisingly, Zak and his colleagues' groundbreaking research on the nature of trust also tells us something about the neurochemistry of distrust. He noted the release of a testosterone derivative in men during moments of distrust and stress. This derivative is called dihydrotestosterone (DHT), and it's no small force to counteract: Elevated levels have been proven to boost the desire for confrontation—often physical—during intensely anxious social situations.

So aside from shooting hormones up the noses of the most powerful men on Wall Street, what could Paulson and his team have done to better establish an atmosphere of trust? If systems create their own behavior, why was Paulson's particular negotiating structure producing such distrust? In essence, how do we work together to solve a vital issue when it is in our collective, but not necessarily our individual, best interest?

One of the greatest challenges to successful cooperation was the complex role that Paulson played. As a former head of Goldman Sachs, Paulson's role as a neutral representative from the government was compromised. To the bankers at 33 Liberty Street, he was still "one of our kind," a member of the investment banking tribe. When push came to shove in the negotiating process, the bankers never really believed that one of their own would allow Lehman, much less the whole system, to fail. Such a belief was not unfounded: Paulson had blinked and caved when he had decided to save Bear Stearns only months before. Surely he would do the same for Lehman. After all, Paulson had been a banker for a lot longer than he had been a regulator. And when his stint in government was over, he might just become a banker again.

The bankers returned for another marathon session of talks on Sunday morning, and the CEOs acknowledged intractable disagreements over Teamwork Saves the Day. When the Barclays representatives turned to Paulson for taxpayer money to help them buy Lehman, he refused. A few hours later, Barclays backed out altogether as a potential buyer. The game was over.

Jim Wilkinson, chief of staff to Treasury Secretary Paulson, spoke to the room as the bankers started getting ready to leave: "This would be extremely interesting from an analytical perspective if it wasn't happening to us."

Briefcases snapped, suits and coats buttoned and zipped: 33 Liberty Street was closing up shop for good. A silence fell over the room as the players wandered out, stunned by a collective sense of dread. One by one, they brushed off the first signs of dust from the fall of the House of Lehman and attempted to focus their fears on how Lehman's collapse would affect their own precarious state of affairs.

When the CEOs left the meeting on Sunday evening, they did not know exactly how the markets would respond to their decision, but they all understood the underlying game play of the weekend: There would be no consensus for collective action.

Over the weekend of September 12, Lehman Brothers filed the largest bankruptcy in American history. Thousands of employees arrived for work on Monday wheeling small suitcases and toting gym bags to empty out their offices. Meanwhile, Merrill Lynch, famously catering to the middle class with its "thundering herd" of brokers, made a frightened dash into the arms of Bank of America for safety.

On Monday September 15, Wall Street had its worst day in seven years. The Dow Jones Industrial Average lost more than 500 points, its steepest point drop since reopening after the September 11 attacks. In one single day, $700 billion disappeared into thin air, leaving retirement plans, government pension funds, and investment portfolios empty. Shares of Morgan Stanley and Goldman Sachs took a nosedive. Over the following days, both companies reconfigured themselves as commercial banks, leaving behind their storied roles as securities firms.

In just one weekend, the decades-long winning streak involving some of Wall Street's greatest gamblers came to a crashing end. On September 15, it was time to walk away from the table and face the harsh light of day. It was every man for himself.

Of course, there were seemingly countless impediments to the success of Teamwork Saves the Day during that ill-fated weekend. We mentioned the air of secrecy and suspicion that hung over the meeting. Not only did that breed mistrust, it also made it impossible for the bankers to recruit other members of their network for support. The lack of transparency did not just apply to their own balance sheets, it was across the entire financial market system. Bad information led to poorly informed decision making and a great deal of uncertainty in the room. It was unclear how effective Paulson was in his role as referee. The bankers' decades of training in negotiating within competitive markets discouraged whole-systems thinking, much like to fishermen using maximum sustainable yield as their default mode rather than ecosystem-based fishery management. And, as if all of the aforementioned were not enough, the terms around the table implied that there might, in fact, be winners among the losers, lending the entire negotiation process the tenor of a zero-sum game.

You would be forgiven for thinking that, considering all of the above, the bankers could not have reached any other conclusion. And yet, the lens of game theory suggests otherwise. The key to how that could be lies in one simple, four-line computer code that has single-handedly changed the entire framework of science's attitude toward cooperation. Its name could not have been more appropriate: Tit for Tat.

TIT FOR TAT

Ever since the age of Darwin, scientists had been puzzling over one seemingly simple but impenetrable question: If living things evolved through competition, how did cooperation ever emerge? In the late 1970s, long before Paulson's tense meetings at 33 Liberty Street, a political scientist at the University of Michigan named Robert Axelrod set himself the task of exploring this vexing question. His research started with the standard paradigm used to measure the evolution of cooperative behavior: a non-zero-sum game called the Prisoner's Dilemma.

The rules of the game sound more like an outtake from *The Godfather* than a mathematical puzzle:

Imagine two burglars, partners in crime, are being interrogated separately for suspected involvement in a jewel heist. Although convinced of their guilt, police do not have sufficient evidence to convict them of the most serious offense but do have enough to convict both prisoners of a lesser charge—let's say six months in prison for loitering outside the jewelry store.

Police, of course, want to convict the criminals of the most serious charges with the longest sentences possible, so they use a classic divide and conquer strategy: In separate rooms, and out of earshot of each other, they put pressure on each man to confess and implicate the other. They make the following offer: "Look, we've got videotape of you hanging around outside the jewelry store. You're already going to jail for six months, minimum. But if you confess to the robbery, and your partner doesn't, we'll let you off scot-free and use your testimony to put that other guy away for ten years. Of course, if you don't confess and your partner does, you'll go to prison for ten years and he'll walk away free as a bird. And if you both confess, we'll put you both away for five years."

Because the dilemma itself can be complicated to explain, it is often rendered in a matrix that looks something like this:

	Prisoner B Stays Silent	Prisoner B Betrays
Prisoner A Stays Silent	Each serves 6 months	Prisoner A: serves 10 years; Prisoner B: goes free
Prisoner A Betrays	Prisoner A: goes free; Prisoner B: serves 10 years	Each serves 5 years

If you put yourself in the prisoners' shoes, you can see where the dilemma part comes in: While it's in both of your interests to cooperate in keeping your mouths shut (following the old Mafia rule of *omertà*) and get a minimal sentence, it's otherwise in your personal interest to defect and rat out your accomplice. The situation is made worse by the fact that you can't communicate or coordinate with your partner. Think how easy it would be if you could just call him up on his mobile phone and make a plan to get you both off with a minimal sentence.

Yet it is this very restriction that allows the Prisoner's Dilemma to mimic our ancient evolutionary landscape. Our mammalian ancestors, after all, had to cooperate long before we evolved the language to ask for it. The game roughly parallels the position of animals, unable to make any promises of a payment and yet fully engaged in cooperation: "You scratch my back, I'll scratch yours."

In short, while the details of the Prisoner's Dilemma sound awfully specific, it functions as an abstract formulation of common situations in which there are optimal payoffs for everyone if they can agree on mutual cooperation (win/win) but strong incentives for each of the two players individually to defect (win/lose).

When Axelrod first became interested in the Prisoner's Dilemma in the 1970s, research into cooperation was limited primarily to science that focused on cooperation within a gene pool. Biologist William Hamilton wowed the world with an elegant equation to explain cooperation that worked in service of gene replication, known as *kin selection*. Axelrod's interest, however, was in an altruism, or cooperation, that was strategic rather than genetic. What would ever make one cooperate if it wasn't an effort to propagate genetic code—either one's own genes or the genes of a family member? Evolutionary biologists, beginning with the brilliant polymath Robert Trivers, began referring to a cooperation based on strategy as "reciprocal altruism."

If the concept of reciprocal altruism strikes you as technical or vague, consider situations in your own life. Most people have, at one time or another, encountered a neighbor or coworker who abided by a principle of "gimme gimme gimme" without ever giving anything back. It doesn't take long for the sensible person to "defect" and cut ties with such an egoist. More difficult, however, is the person in your life who cooperates some of the time and then defects at others. It is harder to stop interacting with such part-time meanies: Remember the Queen Bee in junior high who deigned to speak with you only once in a blue moon, or the annoying friend who paid for dinner only one time out of three. What about your colleague, the guy who goes out drinking every night and then calls in "sick" for half of your group meetings? Do you cooperate again and again, or do you finally defect? And after a defection, when should we forgive and start cooperating again?

In a one-round match of Prisoner's Dilemma, where the players will never meet again, it often makes the most sense to defect immediately, because that is where the payoff is potentially the biggest. Two strangers hailing the same cab have no chance of seeing each other again, so there is little incentive to be accommodating. Most of us, however, live in a world of iterated rounds of the Prisoner's Dilemma, or rounds that involve the same players in the same game time after time after time. Anyone who has ever been married or negotiated living quarters with a partner or roommate intuitively knows the cycle of cooperation and defection that characterizes the iterated rounds. Who took out the garbage last night? Who's going to take it out tonight?

Our forebears existed in small tribes—negotiating over and over again with the same few people—so we have finely evolved mechanisms for strategizing for the longest arm of time. Common sense tells

programs in the competition to make copies of themselves at the end of each round, reproducing in proportion to how well they had done. He then ran the simulation over and over, providing a rough proxy for evolution. Axelrod found that, as long as there was a minimally sufficient number of nice strategies that could meet and reproduce, these strategies could grow and thrive, even in a world dominated by defectors.

Over the long haul, Tit for Tat showed how cooperation could evolve as a stable strategy in an otherwise competitive evolutionary framework. When we take Tit for Tat out of the realm of game theory and into a real-world context, however, the story changes slightly.

Axelrod's computer tournaments focused exclusively on the logic of two abstract dispassionate actors who execute their moves flawlessly. But the real world is nothing like this. People make mistakes—they sometimes cooperate when they mean to defect, and vice versa. And this messiness has important consequences for how the game is played.

For example, when Axelrod studied the results from the first round of his tournament, he discovered that Tit for Tat might have been beaten by a slightly more forgiving strategy, called Tit for Two Tats. As the name implies, this strategy retaliates after only two back-to-back defections by its opponent. It "turns the other cheek," at least initially, before responding.

When Tit for Two Tats competed against another nice strategy like Tit for Tat, it fared marginally better. Yet in another iteration of the computer tournament, in which hostile strategies were also included, Tit for Two Tats was a major loser, finishing twenty-fourth. More aggressive strategies exploited its niceness—defecting, but never enough to trigger retaliation.

Yet, intriguingly, recent research suggests that Tit for Two Tats might actually be a much more successful strategy in the real-world realm of human affairs. To test this, Harvard psychologist David Rand and his colleagues created a modified version of the game. In it, human subjects played a series of repeated Prisoner's Dilemmas, with a twist: In each turn, players had a one-in-eight chance that their move would be changed to the opposite of what they intended—a cooperation would turn into a defection, and vice versa. This condition was added to account for the fact that people aren't perfect—they sometimes make execution errors.

In this error-prone, "noisy" condition, players who followed at Tit for Two Tats actually earned the highest payoff, because they were able to look past the occasional, unintended defection. Players following a stricter Tit for Tat strategy, on the other hand, ended their cooperation too early, and lost out on a beneficial relationship. "In an uncertain world," says Rand, "it can be advantageous to be slow to anger and fast to forgive."

The real-world also varies from Axelrod's simulations in other ways. Human beings are social primates, and like all social primates, our ideas about when to cooperate and when to defect have been strongly influenced by the groups on which we depend for survival. Thinking in social rather than individual terms will often lead us to make counterintuitive choices that are as much about the impact on other members of our group as about the benefits we ourselves receive (and the ones we suspect others are receiving).

Primatologist Frans de Waal, together with his research partner Sarah Brosnan, investigated cooperation and perceptions of inequity among capuchin monkeys. In one of their experiments, two monkeys were given either a grape or a piece of cucumber for completing a simple task. If both monkeys were given the same reward, de Waal and Brosnan didn't observe any problems. They confirmed that the grapes were far preferred—de Waal noted that all primates enjoy a sugar rush—but even if both monkeys received a cucumber, they seemed to have no problem repeatedly performing the given task.

The experiment started to get interesting, however, when monkeys received *different* rewards. The one who got the lesser reward—a cucumber instead of a grape—would start to hesitate, eventually putting up a fight by either not eating the cucumber or not performing the task.

De Waal concluded that this is an irrational response: "If

profit-maximizing is what life (and economics) is about, one should always take what one can get. Monkeys will always accept and eat a piece of cucumber whenever we give it to them, but apparently not when their partner is getting a better deal."

Humans have this same instinct—referred to as *inequity aversion*—but economists, game theorists, and mathematicians have historically counted on us to care more about the pursuit of self-interest over fairness. Yet when it comes to the distribution of scarce resources, there is little evidence that humans actually do act purely in accordance with their own self-interest. Even when we have a fresh, tasty cucumber in our hands, humans have evolved the tendency to start fixating on the asshole with the grape.

All of this is to say that, despite the dictates of Adam Smith or Milton Friedman, the bankers at 33 Liberty Street were still using their social and emotional brains. Amidst discussion surrounding systemic risk and toxic debts, every one of the suits around the table was also assessing who would get stuck with the cucumber and who would walk away with the grape.

In addition, the bankers also brought a whole set of cognitive biases that help to inform how all of us, as social creatures, make decisions. Each CEO had worked directly with a team of hundreds, perhaps thousands, of bankers and had met many more in the course of his career. Yet this is just a small fraction of the whole—in the global world of finance, there are many thousands of players; even a CEO in an alpha position such as Mack, Blankfein, or Thain couldn't possibly know every single banker in the field. These CEOs—like the rest of us—use heuristics (mental shorthand) to identify colleagues in different firms, akin to different tribes: "You know Bob; he's with Morgan Stanley." These tribes in turn have their own values, cultures, myths, risk tolerances, and styles, which help bind them together and reinforce distinctions for their members. And tribal thinking can assert itself with a vengeance.

As the bankers considered bailing out Lehman, for example, they were threatened by the thought of handing over their scarce resources to help Barclays sweep in and score with a strategic and profitable move. Barclays registered in mental shorthand not just as "threatening" but also as "British." In cognitive terms, when negotiating over possible money coming in from the American taxpayers, Barclays was an outside group—a foreign tribe. The *Wall Street Journal* noted that the bankers "were loath to provide support when a rival like Barclays might still buy Lehman." Later they reported, "By Sunday morning, the U.K.'s Barclays looked like the sole potential buyer. That further minimized the chances of a government bailout: If the Bush administration wouldn't help to fund a Wall Street solution, aiding a foreign buyer was even less likely."

Is it possible that oxytocin—our love and trust elixir—is the very same hormone underlying such in-group and out-group biases? Scientists like Carsten K. W. de Dreu, a psychologist at the University of Amsterdam, believe so. He gave Dutch students a standard moral dilemma: Do you save five people in the path of a train by throwing one bystander over the tracks? He left the five people nameless while giving the sacrificial victim either a Dutch or a Muslim name. Dutch subjects given a whiff of the oxytocin before beginning the experiment were far more likely to push Muhammad over the tracks than Maarten.

A famous study from 1954 suggests similar hormones at work: Psychologist Muzafer Sherif and his colleagues gathered twelve-year-old boys at a summer camp, divided them up into two groups, and instructed them to compete against each other. The boys quickly started to antagonize members of the other group, doing things like holding their noses in disgust upon seeing them pass by—behavior known as *out-group derogation*. Meanwhile, within the groups, cohesiveness increased, and the groups almost immediately became both more hierarchical and more internally cooperative.

This kind of in-group, out-group tribalism has a powerful logic. Game theorists Steve Rytina and David L. Morgan mathematically modeled this bright line between "Nice Nellies" and "Meanies" by dividing an imaginary society into two groups, the Blues and the Reds. They watched what happened when members of each tribe followed a variant called Discriminatory Tit for Tat (DTFT), which is just like Tit for Tat except when dealing with someone of a different-color group. A Red will always defect with a Blue, and vice versa.

When two Reds engage in negotiation for the first time, each cooperates. When two Blues with no history negotiate, both cooperate. But what happens when a Red and a Blue interact? Defection is automatic because "you can't trust those guys." Think of the Sharks and Jets; the Montagues and the Capulets; the Hatfields and the McCoys. Unfortunately, we understand this story all too well.

Rytina and Morgan demonstrated that the DTFT game play is not only stable, it's all but intractable. In the early rounds, an individual who tries to play regular, color-blind Tit for Tat is *worse* off than one who defects with the out-group. Why? When a Red and Blue interact for the first time, even if the Blue contemplates cooperating (as in regular Tit for Tat), the Red player will almost certainly be playing DTFT and will defect. That means that the Blue player—only trying to start off nice—will get the sucker payoff and lose points.

This does not mean that DTFT is a more successful strategy. Tit for Tat still leads in terms of long-term resilience. The problem with DTFT is in its insidious stability. Once it is entrenched, it punishes individual efforts to attempt cooperation across enemy lines.

So what is the takeaway of such results? As in any complex system, it's a mixed bag. The bad news is that we have evolved cognitive biases that lead us to see our playing field in terms of "us" and "them." The good news is that, although this is a stable game play, it has been mathematically proven to be less successful (over the longest time periods) than the more generous strategies of Tit for Tat.

But considering the rational and emotional rewards we get from aligning ourselves with a tribe (DTFT style), how do we trigger cooperation between opposing groups?

Fortunately, there is a surprising amount of flexibility within de-lineations of in-groups and out-groups. As the famous sociobiologist E. O. Wilson has written, "Altruism is characterized by strong emotion and protean allegiance. Human beings are consistent in their codes of honor but endlessly fickle with reference to whom the codes apply." Note the use of the word "fickle." It's our very plasticity that gives us our evolutionary advantage. Yes, we love "our" people. But our minds have evolved malleable markers for the definition of just who "our" people are. Wilson goes on to write, "The important distinction is . . . between the in-group and the out-group, but the precise location of the dividing line is shifted back and forth with ease."

The trick, then, is to get people to expand their definition of "us." But how?

ENLARGING THE TRIBE

Arthur Aron and Elaine Aron are husband and wife as well as research partners and writing collaborators at Stony Brook University. They have made it their life's work to study the ways in which we form intimate ties with those in our inner circle as well as with members from an out-group. Over the last decade, they have been bringing together diverse pairs of complete strangers—black and white, Latino and Asian, black and Latino—to participate in an unusual experiment.

The Arons asked each of these couples to come together for a series of four intimate, hour-long sessions. During their first session, each member of the pair was asked to share his or her answers to a list of questions, everything from, "Would you like to be famous? In what way?" to "If you could change anything about the way you were raised, what would it be?" During the second session, the pairs competed against other pairs in timed games like charades, word play, and logic puzzles. Then, in the third, they were guided through a series of inti-mate conversations with questions about their personal lives and their feelings of affiliation toward their ethnic groups.

In the last hour-long session, they did a blindfolded trust walk, taking turns navigating a maze wearing the blindfold or serving as the guide.

Although the activities conjure up a weekend at seventh-grade summer camp, Dr. Arthur Aron argues that these four hours create a relationship that is as close as any in a person's life. In essence, they are saying that these relationships, when successful, are just as effective as any at releasing the hormone oxytocin. Once the "cuddle chemical" is triggered—whether through the structure of the Arons' exercises or a whiff of Zak's nasal spray—trust and cooperation naturally follow.

In fact, research from the Arons' tests showed that the four-hour sessions almost immediately lowered subjects' score on a variety of prejudice measures. Stress hormone tests, conducted on the subjects' saliva, showed significantly reduced anxiety for both members of the pair when they encountered a social interaction with a member of their partner's ethnic group.

Psychologists suggest that these kinds of powerful bonds—creating new definitions for in-group and out-group—occur through our own highly evolved process of empathetic mimicry. We are rarely conscious of it but we, as humans, are constantly copying the facial expressions, manners of speech, postures, and body language of those around us.

All of these copying behaviors, through a rather elegant neural feedback loop, allow us to actually experience the emotions associated with the particular sort of behavior we are mimicking. This helps to explain why it is so excruciating to watch someone else cry: Your neural feedback is allowing you to feel an analogous sensation. It makes human tears—costly in energy for the body to produce—a bona fide evolutionary bargain when it comes to the empathic and cooperative payoff.

It's no coincidence that the Arons take their couples on a walk, the quintessential exercise for conflict resolution. In 1982, when arms negotiators from both the United States and the Soviet Union were looking for a way to draft an arms reduction proposal, they met in Switzerland to meander on that now famous walk in the woods. Face-to-face interactions often bring conflict to the forefront of the conversation, whereas walking side by side allows space for a more considered dialogue. What if we could take not just a few pairs, but hundreds, and then thousands of people on a trust walk together? Could a walking path help to enlarge the tribe in the midst of one of the most fraught tribal conflicts on the planet?

William Ury, a world-renowned expert on conflict negotiation, was committed to finding a way to bring religious storytelling into the sphere of political discussion in Middle East conflicts. Despite his stature in the field of conflict resolution—Ury is cofounder of Harvard's Program on Negotiation—most of his colleagues counseled him to let go of religion as any kind of negotiating tool. But throughout the Oslo talks in 2000, Ury was dismayed to see that the political process failed to address the two aspects of the conflict that he deemed essential: an acknowledgment of the identity issues wrapped up in the distribution of the land and the practical, economic conditions of the Palestinians (in decline since the 1990s).

"Things were so stuck that you really needed out-of-the-box thinking. The question was and is: What is the game-changing move?"

An anthropologist by training, Ury looked to some of his work in other cultures for answers. He realized that the conflict needed a way to call upon the community he described as the "third side."

"Spending time with groups like the Bushmen in the Kalahari, I noticed how they used the surrounding community as a third side. When they approach conflict, everyone gets in a circle and they all have a say. They play an enormously healing role in which they create a container. Even a destructive conflict can be gradually transformed when it is contained inside community." In other words, when everyone is a member of the same tribe, the oxytocin rush of pitting us against them disappears.

Ury was looking for a set of people to act as a buffering force. But who was the third side in the Middle East?

"Every culture has an origin story," Ury told us, "and the origin story in the Middle East engages almost all of us on some level. Four thousand years ago a man and his family walked across the Middle East. The world has never been the same since."

That man, of course, was Abraham.

Ury started referring to him as our symbolic third side.

"What if you took the story of Abraham—the symbol of hospitality—and used it as the antidote to terrorism, a vaccine of sorts against intolerance?" By calling upon the story of Abraham, Ury could unite the three groups of faith and, at the extreme fringes, nullify the faiths in relation to one another. All three faiths defined under one tribe—People of the Book—had a shared history, shared values, and a shared sense of God that conferred respect on everyone.

"In most negotiations I've been in, there is something so simple that is required. I would call it respect. It's the cheapest concession you can give as a negotiator—it doesn't cost anything—but it's amazing how little we use it. Respect is an interesting element because it's a positive-sum element. If I give you respect, I don't have less respect for myself. In fact, you're likely to give me more. That is what Abraham actually stands for: respect and the positive value of the other human being."

But it wasn't enough for people simply to remember or recite the story of Abraham, Ury realized. For the symbolic third side to really play a role in the conflict, people would need to experience the story. And the best way to do that would be to go for a walk.

"Walking has power; it's what made us human. A path that followed in the footsteps of Abraham—crossing ten different countries—has the power to transform hostility to hospitality, terrorism to tourism, all in the name of Abraham."

Ury decided to trace the journey of Abraham through the Middle East and use it to create a cultural, ecotourist walking path. He named it the Abraham Path Initiative.

The route Ury and his colleagues originally followed started in Abraham's birthplace in Urfa, Turkey, then wound through Harran, Turkey—the place where many sources suggest Abraham heard the call from God. It continued onto Syria, down from Jordan to Jerusalem, and then into the West Bank before ending in Hebron, or al-Khalil, described in the book of Genesis as Abraham's burial place. Ury describes this trek—more than six hundred miles—as a journey from "womb to tomb."

Just as the Arons surmised in their experiments on intimacy, the simple act of walking provides an opportunity for participants to engage in small acts of trust building. These interactions build opportunities for hospitality: the cooperative contract between guest and host based on mutual respect.

"In the West," Ury told us with his characteristic wry chuckle, "we perceive the conflict through a frame of HOSTILITY. But you take HOSTILITY and add PITA bread, and you get HOSPITALITY."

And that is exactly what we discovered when we donned our hiking boots and experienced the Abraham Path for ourselves. Our five days of walking through the shepherds' hills of Palestine took in but a small section of William Ury's game-changing vision, but they put us—and our fellow walkers—in touch with dozens and dozens of new faces as well as frame-changing experiences.

Not only were we served fresh pita bread drizzled with olive oil—made daily in the caked mud ovens at the back of the homes—we also encountered an almost dizzying array of responses toward religion, the role of women in society, and the identity of the modern Palestinian. We were welcomed in villages where the men refused to shake the hand of a woman, much less sit in a room with her. But we also sat and drank sweet tea in villages where the women functioned as powerful matriarchs—outspoken and well educated—while the men sat back in respect. The one constant was the contract of hospitality—respect—spoken in the name of Abraham.

When we walked up the steep hills of Awarta, five miles southeast of Nablus, we were even pulled to participate in a town wedding without anyone ever batting an eye. There, in a meeting house in the center of town, hundreds of traditionally dressed women from the village were dancing and singing with abandon while a young bride and groom sat elevated on chairs, watching them. The singing—referred to in Arabic as *al-mardudeh*—was a raucous call and response between the lead singer, known as the *badda'a* (the talented one) and the hundreds and hundreds of women all packed into the tiny room. Before we had time to fully enter the room, we were placed next to the bride and the groom on the elevated seats overlooking the crowd of celebrants.

"You are special guests here in the village," a woman wet with sweat told us in between claps and chants. "We must celebrate you!"

Not only invited to a complete stranger's wedding but deemed a guest of honor: It is a gesture of such generous hosting that it is almost inconceivable in most Western countries. Frédéric Masson, the French-born coordinator for the Palestinian arm of the Abraham Path, explains that this extraordinary culture of hospitality is being threatened by the lack of employment and social structure in the occupied territories.

"With the current situation, the villagers are isolated from the world. They are so eager to share their experiences but they are disconnected and that leads to a great despair, especially for a culture that values hospitality. When you attended their wedding, you were honoring them and their expression of hospitality."

This leads to the second essential aspect of peacemaking that Ury is eager to address: the idea of mutual prosperity, another non-zero-sum element. The tourism along the path brings in income for the villagers: selling food, teaching classes, offering shelter. Their culture will be valued in name but also through economic improvements, conferring even more respect on the people offering up their hospitality.

Over the last five years, thousands of people have started to walk the path through Syria, Jordan, Israel, and Palestine. And people unable to journey to the Middle East are organizing walks in their own cities and communities. Much like the layered garden of Willie Smits, the Abraham Path attempts to address both immediate economic needs of the community as well as its needs for respect and trust over the longest arm of time.

"Today the path is like an acorn," he told us. "Tomorrow it may look like an oak tree."

His words remind us that though Abraham serves as the symbolic third side, we, the global community, are the ones who will walk the walk.

Now think back to our bankers at 33 Liberty Street. As we have already established, the secrecy of their meeting impeded any efforts to enlarge their tribe. The nature of the negotiations created strong incentives for side dealing, as the participants engaged in two very different kinds of behavior. It is hard to think in zero-sum and non-zero-sum terms simultaneously. Worst of all, the bankers—even with Paulson in the room—represented only a tiny fraction of the real stakeholders in the crisis. What might have changed had Ury's third side—ordinary citizens, other kinds of companies, outside economists, or activists—been present to enlarge the tribe and seek out a constructive whole-systems solution?

Despite all the positive evidence from neuroscience and game theory, and the demonstrated power of strategies like enlarging the tribe and employing the third side in complex negotiations, one might still doubt that the bankers at 33 Liberty Street could ever overcome the seemingly insurmountable odds against them and find a path to successful cooperation.

And yet, less than two years after their fateful meeting, in the middle of an even more profound crisis, with a deadline similarly measured in hours, and with life and death hanging in the balance, a group of total strangers showed just how such a cooperative effort could emerge. Their story—of Mission 4636—holds lasting lessons for eliciting cooperation everywhere.

MISSION 4636

The devastating earthquake near Port-au-Prince, Haiti, on January 10, 2010, wasn't the most powerful earthquake ever to hit the poorest nation in the Western Hemisphere, but it was certainly the most catastrophic. As the quake ripped through the ground below, it triggered the collapse of nested socioeconomic, political, racial, and physical fragilities above. In mere moments, it generated statistics of the sort one normally hears tallied after a war: 316,000 Haitians were killed, and another 300,000 were injured. A million more were made homeless; in all, three million were affected. Physical infrastructure was leveled, entire departments of the government were obliterated, the citizenry was traumatized, as were countless international aid organizations who were already operating in the country. There was never any doubt that this would be a humanitarian crisis of the first order.

And yet, as terrible as the days' events were, the earthquake might have been an even greater disaster had it happened just a few decades earlier. For on January 10, 2010, the instantaneous news of the disaster also unlocked a uniquely collaborative, global response that could not have existed before the age of the Internet—one that harnessed a global community of volunteers, technologists, first responders, and Haitian citizens both at home and abroad.

What's more, this remarkable collaboration unfolded in approximately the same amount of time as the meetings held over that now-famous weekend at 33 Liberty Street.

Our story starts 1,700 miles away from Haiti in the snowy winter weather of Somerville, a suburb of Boston. Rosalind Sewell—known as Roz—had just returned to the East Coast to resume her first year of graduate school at the Tufts Fletcher School of Law and Diplomacy when she heard the news of the quake. As a former Fulbright fellow in Morocco, Sewell was already interested in emerging forms of social media and technology around the globe. That's why, when she checked her email the week of the tenth, one message in particular caught her eye.

"The email said, 'Hey, if you're in town, we're sitting around in my apartment mapping social media as a response to the earthquake.'"

The man behind the email invitation was a PhD candidate at the Fletcher School named Patrick Meier. Sewell recognized him as the leader of a group of Tufts students particularly interested in crisis mapping, a new field of disaster analysis that used a combination of satellite imagery (now widely available online from organizations like Google) and data sent in through SMS texts, Twitter, and Facebook to map out the impacts of an unfolding crisis and coordinate response.

For Sewell, the email promised an opportunity to learn more about crisis mapping and the community that Meier had cultivated at Tufts. When she showed up at Meier's apartment that evening, she expected to see a group of Tufts students calmly reading Twitter and Facebook and then uploading the coordinate information to an online map. What she encountered, instead, was a full-blown operations center, the nerve center for information sharing and incident reporting on the Haitian earthquake.

"My family is army, so I've seen the Army Op Center with the thousand computer screens and the big monitors. When I walked into Patrick's living room, it had that exact feel. A couple of people were running around on telephones, putting information in, really focused. Someone else was trying to train new people. The whole apartment had this really electrifying energy that sucked me in."

She arrived that night planning to map for a couple hours and then return home. Instead, she stayed inextricably linked to the crisis mapping effort—both in person and online—around the clock for the next three weeks. And Roz Sewell was not alone. Thousands of people all over the globe—from Istanbul to Geneva to Washington, D.C.—collaborated in response to the earthquake, making it the largest crowd-sourced crisis mapping effort ever undertaken.

Mission 4636 actually had its roots halfway around the world in response to another crisis: a terrifying wave of postelection violence that had buffeted Kenya in 2007. In the wake of the election, ethnic violence swept through the countryside—a cycle of horrible defections among groups: Discriminatory Tit for Tat, gone terribly wrong.

In the midst of the crisis, Kenyan residents had virtually no reliable real-time information on the violence outside of the cities. Ory

Okolloh, a Kenyan-born Harvard-trained lawyer, who had returned to Kenya to observe the elections, started to collect and geotag all the incidents of violence on her own personal blog, but the reports were coming in too quickly. Much like financial regulators before 2008, Kenyans needed a better mapping tool, one that would dynamically change to reflect real-time information on violent outbreaks. But they also needed a system they could access themselves, and easily. She reached out to the blogosphere with a challenge to help her create such a tool. Days later, volunteer software developers answered her call by quickly prototyping a web platform that allowed Kenyans to contribute anonymous reports of riots, rapes, deaths, and displacements of people using just their cell phones. These crowdsourced crisis reports could be mapped on a single display, and citizens could then access them not only to see what had happened and where, but to see larger trends over time—how the violence had spread through the community and where it might go next. The loose network of technologists and human rights activists called the tool they had built Ushahidi, which means "to bear witness" in Swahili.

In the immediate circumstances of the election, not enough people were aware of the program to be able to make it truly usable during the crisis, so the tool served as more of a prototype, a vision for what might be. One thing, however, was clear to the developers right away: The platform could be used by a community in any situation that was lacking good, real-time systemic information—not just a violence outbreak, but things like public health epidemics and ecological disaster. They decided to package it for a wider variety of contexts.

Over the last few years, the Ushahidi platform has been used by communities to track anti-immigrant violence in South Africa, map violence in eastern Congo, report pharmacy stockouts in several East African countries, and to monitor elections across the world. The organizational goal of Ushahidi was to make the platform easy to use and free of charge to anyone who wanted to bring greater transparency to an issue. Like many software developers, the organization that created

Ushahidi never intended to get involved in any of the actual deployments after their initial launch in Kenya.

On January 10, all of that changed.

Patrick Meier, the man behind the email sent to Roz Sewell's account, is the director of crisis mapping at Ushahidi and the cofounder of the International Network of Crisis Mappers. He was watching CNN in Boston a few hours after the earthquake struck, shocked by the severity of the damage. Meier called Ushahidi's main technical director and told him that he really wanted Ushahidi to do something beyond just disseminating the platform. Considering the extremity of the disaster and the impoverishment of the country, Meier felt that the quake was set to be a tipping point—a systemic flip—for the whole country.

Within an hour of getting Meier's call, Ushahidi's technical director worked with its main programmer to customize the platform for the Haitian crisis. Whereas the Ushahidi platform was used in Kenya solely as a mapping visualization tool, in the Haitian disaster, it was hoped it could also be used to help coordinate responses, connecting all of the stakeholders in the crisis: first responders, NGOs, the government, and most important, Haitians themselves. This was a fundamental shift in the way the platform functioned: It went from being simply a mapping tool to a tool for intervention.

"Once I made the call to set up the platform, it was clear that there was no way Ushahidi's small, primarily tech team was going to be able to do the near real-time crisis mapping," Meier told us. "After Kenya, this is the only other time we've ever done a deployment."

The first requirement was to track down and sort all the information streams coming out of Haiti, including tweets, Facebook updates, people finders on various sites all across the web, and any text messages that were managing to make their way out of the country via mobile phones.

Meier knew that monitoring all of this information would require the constant attention of a whole team of people. But that was only the beginning. After tracking all the pieces of information, they would need to be processed into different categories and then geotagged, mapped onto a dynamic digital map of Greater Port-au-Prince with accurate coordinates. Considering the fact that neither Meier nor any of the other immediate Ushahidi team was familiar with the streets, back alleys, and local haunts of Port-au-Prince, the task of finding the right coordinates seemed next to impossible.

To aid in the seemingly insurmountable effort, Meier called upon Ushahidi's latent network, sourcing potential volunteers through Twitter, Tufts University, college campus networks, and the social circles of a generation of digitally savvy development professionals. The Ushahidi team was also part of a large and informally linked patchwork of likeminded, technologically innovative humanitarian NGOs, technologists, and researchers that had been building steadily for years, many of which, at various points, would play important collaborative roles in the Haitian crisis response.

The SOS email that Roz Sewell received went out to the thousands of other people on the Tufts University email list. As an affiliate of the Fletcher program, Meier had already given some presentations on Ushahidi at the school, galvanizing a small group of friends and acquaintances engaged in the effort. After his initial email, the core group of a dozen volunteers showed up, accompanied by a few new faces like Sewell. Meier trained his initial group on the basics of tracking and tagging, emphasizing the urgency of the effort while giving them ideas and tools for finding more people to help. He told them that the work was remote: As long as they signed into the group Skype chats and logins, they could process and map information from anywhere in the world.

And then he left them alone.

After the initial handful of people showed up to help in Meier's living room, the numbers of volunteers grew quickly. A few days later, twenty people were squeezing onto the couches and floor space. Then thirty, forty, and, finally, when more than fifty people were coming to help, the group arranged to take over a study space in a Tufts University building. By the following week, a training session arranged for Tufts undergraduates was overflowing with more than eighty students and the op center was given an official name: Ushahidi-Haiti @ Tufts.

Soon Meier was receiving email notifications that training sessions and ancillary crisis mapping situation rooms were popping up all over the world. A group of Tufts alumni emailed Meier and told him that they were replicating the crisis mapping op room down in D.C. with login information passed on from another volunteer. An exchange student from the Fletcher School studying abroad at the Graduate Institute in Geneva emailed Meier and the group and told them that she was covering Geneva with crisis mapping outposts. Another Fletcher student still in Portland for the holiday break started organizing students to do real-time mapping from the West Coast. Then the seeds of organization blew to cities all over the world, creating small start-up communities of crisis mappers in London, Montreal, Providence, Istanbul, and other locations across the globe. Anywhere there was at least one person with the training to log in to the Skype chat and the platform, there was a crisis mapping node.

Just like the previously discussed, malevolent AQAP networks, these pro-social crisis mappers were not bound together via traditionally strong command-and-control structures, but by informal social connections and a shared set of values. Over time, hundreds of crisis mappers were contributing to the effort—a fully fledged swarm. Meier had not met a single one of them. In some cases, he was completely unaware they existed until the entire effort was over.

"I told people, 'We're going to let this be emergent,'" Meier explained. "There are so many things that need to happen every single hour and so many things that need to keep evolving in such a short amount of time. I have to just let it flourish and deal with what happens when it starts getting inefficient."

The open nature of the platform—both the code that powers Ushahidi and the collaborative nature of the mapping—meant that people could easily be recruited to perform discrete, useful tasks with a minimum of formal authority. That helped the platform, and the recruitment process, to spread.

"The whole story of the volunteers is one of the most amazing things that I've ever been a part of," Meier reflected. "Within about five or six weeks, people trained more than three hundred individuals on how to use Ushahidi's platform, and I didn't organize a single one of the sessions."

In the earliest hours of the crisis, this ad hoc global crisis mapping community was pulling its data primarily from news reports, Twitter feeds, and Facebook updates. But to truly transform from a mapping tool to a full-blown intervention, Ushahidi would need first to directly connect with Haitians in need, and then in turn connect them with the first responders (who could look at the map and determine where they could be the most helpful). While the crisis mappers were in the midst of recruiting and training volunteers to assist in the deployment, another initiative, led by a constellation of technology-led NGOs including FrontlineSMS, InSTEDD, Energy for Opportunity, and the Thompson Reuters Foundation, was under way to figure out how to enable Haitians on the ground to add in their own situation reports to the map.

On January 10, minutes after the earthquake hit, Josh Nesbit, a self-described health-care junkie in his midtwenties who worked for FrontlineSMS (now MedicMobile), made a call to the State Department. Within an hour, he was strategizing with media-savvy staffers there about using radio and the telcos to reach the Haitians directly.

"I basically knew that in a country like Haiti where only about one percent of the population is online—even less than that after the quake—land lines are pretty much nonexistent," Nesbit told us. "On the other hand, forty percent of the population owns a cell phone, and if you look at people who have access to those phones, the number is probably much higher, closer to seventy-five or eighty percent."

Nesbit and his contacts at the State Department knew that the success of the effort would hinge on the ability of the mapping platform to directly connect with Haitians on the ground. They all agreed that securing an SMS channel was a first priority.

SMS—or short message service—is really just a technical term for a text message. What Nesbit and the State Department wanted, however, was a specific channel for receiving incident reports from earthquake victims. Much as cities often set aside phone numbers for specific functions—in New York, people know they can dial 311 to reach the offices of the city government—Nesbit wanted to get all the Haitian telcos to agree on an earthquake emergency text number.

From his apartment in Washington, D.C., he cataloged all the nongovernmental organizations he had worked with over the past six months and pinged them all, looking for what he referred to as an "SMS hub." He spent the rest of the night connected, circling round and round phone calls, tweets, and texts with the team at State and his own personal network of global health contacts. In the flurry of communication, Nesbit rushed out a quick tweet to his entire network of followers:

Reaching out to @FrontlineSMS users in #Haiti with hopes of establishing local SMS gateway for http://haiti.ushahidi.com.

Almost immediately, Nesbit received contact from one of his Twitter followers in Cameroon in West Africa, telling him to reach out to Jean-Marc Castera, Digicel's IP manager in Haiti. Within minutes, Nesbit was Skyping with Castera, making a plan for creating a short code.

"Now, if anyone ever asks me why I 'waste my time tweeting,'" Nesbit told us, "I have a pretty good response for them."

Finding Jean-Marc Castera was like hitting the jackpot for Mission 4636. Instead of approaching Nesbit's request with suspicion or reservations, Castera immediately agreed to get in touch with the CEO of Digicel, Haiti's largest mobile phone telco. What he came back with was better than anyone could have imagined. He proposed that the entire disaster relief mission co-opt the SMS code 4636. Under normal circumstances, 4636 was a simple service providing weather information through an auto reply. Now it could function as an SMS hub for incoming disaster-related messages from Haitians in distress.

Turning that technical possibility into a reality, however, would prove a daunting, on-the-ground challenge, and here the effort saw new, largely unsung heroes enter the story: Eric Rasmussen and Nicholás di Tada from an organization called InSTEDD, a nonprofit organization dedicated to innovating new responses to diseases, disasters, and emergencies.

A year before the earthquake, the Thomson Reuters Foundation, based in the United Kingdom, had commissioned InSTEDD to develop the Emergency Information Service, a web- and mobile-based platform that could be used by journalists to help them, and survivors, better communicate in a disaster. Within hours of the Haitian earthquake, the foundation asked InSTEDD to fly to Haiti and set up operations. Sixty hours after the first tremor, Rasmussen and di Tada and several foundation staffers landed and set up operations at the east runway of Port au Prince's international airfield. It was one of the first organizations of any kind on the ground after the quake.

From their base, the InSTEDD team was able to do several critically important things: First, they built and connected the underlying infrastructure for receiving 4636 text messages, working with both of the dominant telecoms in Haiti (Digicel and another company, Comcel) to ensure that 90 percent of Haitian mobile users could send in texts to the code. Next, they worked with Ushahidi to ensure that they and other groups could get access to the messages being sent in to the system. Again, the latent network of connections paid off handsomely: At the time of the earthquake, InSTEDD's Vice President of Engineering, Ed Jezierski, was on site with Ushahidi in Nairobi and was able to ensure smooth integration between the two platforms.

Then, InSTEDD and Thomson Reuters Foundation teams traveled to radio stations all over Port-au-Prince and got the emergency short code announced over the airwaves. For all of its technical novelty, Mission 4636 would have been irrelevant without the old-school platform of radio: Everyone listening to the radio heard the message about 4636 and they helped to spread the word to friends and neighbors. Haitians hurt, trapped, or in need of food and water could text their message and geo-coordinates to 4636 and the information would be instantly transmitted and integrated into the dynamic map by crisis mappers.

Two days after the quake, most of the technical platforms were in place, the recruitment process for the crisis mapping volunteers was gaining momentum, and a short code had been secured. There was still one enormous obstacle to success, however: Almost all of the victims' messages would be written in Haitian Kreyol—and almost all of the crisis mappers and first responders spoke English. Somehow Mission 4636 needed to find a group of people fluent in both Haitian Kreyol and English who could serve as translators. In the spirit of emergent cooperation, a most unlikely candidate stepped forward to take on the job.

A year earlier, while Nesbit was working for his own health-care start-up, Frontline SMS, in Malawi, he connected with a computer linguist named Robert Munro. After leaving Malawi, Munro returned to Stanford University to continue his doctorate research investigating ways to process large volumes of multilingual SMS messages. Twenty-four hours after the quake, Munro was taking a train from Palo Alto to his home in San Francisco. He checked his email en route and saw a new message from Nesbit.

"Josh asked me if I could adapt the SMS classification work that I'd been doing with him in Malawi for Haiti."

By the time Munro arrived back at his apartment, he was already logged in on several email chains and Skype chats. He would spend every waking hour on Mission 4636 for the next ten weeks.

At first, he was deeply involved in the technical discussion with Ushahidi about how best to process the text messages that would filter in through 4636.

"As we scrambled to launch the technical platform," Munro told us, "we realized that we didn't have anyone to process the messages. I speak some Sierra Leonean Krio, and some French, but not enough that I was comfortable translating messages for many weeks. None of us had really slept more than an hour or so in the past three days, and it felt like everything I was coordinating was about to fail."

Munro realized that he would have to take charge not only of finding translators but also of training and guiding them throughout the effort. Considering his lack of experience in management, this was a daunting realization, but the collective energy and momentum of the mission made it inconceivable to give up. He forged ahead, looking for ways to connect with Haitian Kreyol speakers.

Considering the minute-by-minute contact the 4636 team had with one another via Skype chats, one imagines that they all knew each other well, or that they were, at the very least, in the same time zone. The reality was just the opposite: Many of the individual players had never met in person, and many of the new recruits were communicating via the Skype chats from all over the world. While Munro tackled the task of finding Kreyol speakers also fluent in English, Nesbit was helping new partners plug into the platform.

"The ability to explain the system and the partners involved ended up being hugely important," Nesbit told us. It is no coincidence that none of the technology from Frontline SMS—Nesbit's organization—was in use with Mission 4636. "Having an independent connector with no skin in the game fostered trust much faster."

Back in his apartment in San Francisco, Munro finally found a solution for contacting potential translators by using Facebook. The Haitian diaspora had started to cluster around several groups on the site, and many of them expressed a strong desire to offer their services. Munro was soon coordinating clusters of volunteering groups all over the world including Union Haiti, a group of Haitian expats in Montreal, and members of the Service Employees International Union (SEIU) in the United States, who worked tirelessly in long shifts to translate the texts. As a group familiar with the streets and signposts of Port-au-Prince, their tactical knowledge proved invaluable to the effort.

Within mere days after the quake, the entire system was in place and functioning. Radio broadcasts all over Haiti announced 4636 as an emergency short code. Haitians—many in desperate need—used their cell phones to send text messages into the 4636 SMS hub. These messages were written in Haitian Kreyol with street names and addresses that would make sense only to native residents of Port-au-Prince. The information was then instantly transmitted to Haitian Kreyol speakers all over the world who quickly translated the messages into English while using their on-the-ground knowledge of the city to find exact geographic coordinates on the map. This English information was then instantaneously sent to crisis mappers working around the world, who took each one of the translated messages and geotagged them on the real-time map. This entire process, spanning the entire globe and thousands of volunteers recruited from different organizations, universities, and companies, took approximately ten minutes in the earliest stages of the effort and less than two minutes toward the end.

The last and final goal—to connect these mapping efforts to the community of first responders—came when Ushahidi was linked to UNDAC (Disaster Assessment and Coordination), which directed search and rescue teams to the addresses with the greatest need. Within twelve hours, the Ushahidi map had become an important resource for many first responders from the United Nations, the U.S. Southern Command (SOUTHCOM), and the Marine Corps.

Together, the 4636 team, along with thousands of strangers from all over the globe, had built a pioneering digital disaster response system, largely from scratch—without a single organization or person in charge.

A testimonial on Mission 4636's blog speaks to the effectiveness of the commitment exhibited by Meier and others:

"I cannot overemphasize to you what the work of the Ushahidi/ Haiti has provided. It is saving lives every day," Craig Clark, open source intelligence analyst for the Marines, wrote in during the earliest stages of phase one. "I wish I had time to document to you every example, but there are too many and our operation is moving too fast. . . . The Marine Corps is using your project every second of the day to get aid and assistance to the people that need it most. . . . Keep up the good work! You are making the biggest difference of anything I have seen out there in the open source world."

Several days into the effort, Mission 4636 started shifting from phase one to phase two. According to Nesbit, "There weren't a lot of people to be pulled out of rubble anymore. We moved to more general needs like food and water requests and medical services."

The U.S. Coast Guard, along with Southern Command, became the primary response team during this second phase, helping to connect the victims with food, water, and health services. And an increasing number of organizations were using the maps to plan and coordinate relief efforts.

The Kreyol-speaking translators came together online to share maps and information, bringing members of the Haitian diaspora deeper into the information loop. Eventually, the job of translating the messages was turned over to Haitians still residing in Haiti. Through the collaborative support of CrowdFlower, a private company that develops technical platforms for crowdsourcing, and Samasource, a nonprofit specializing in microwork digital projects, the translation tasks were successfully passed off to paid workers.

There are no official statistics about how many lives were saved using Mission 4636, but CrowdFlower reports that the average response time to translate, map/geocode, and categorize a message never exceeded two minutes. In total, they processed more than 100,000 SMS messages and, at peak volume, more than 5,000 SMS messages were processed in one hour.

By the time Mission 4636 was finished, a whole ecosystem of collaboration—coders and software architects, volunteer mappers, Haitian American translators, Haitian citizens, NGOs, first responders from the Marines, the Red Cross, the State Department, and the UN all worked together on it—most without once having met one another in person.

TRIBES, NETWORKS, AND TEAMS

What made a collaborative effort like Mission 4636 work against such long odds? A big part of the story rests in the structure of the social network that tied together the volunteers.

Social networks are classically described by being comprised of weak and strong ties. Your relationship with a close friend or family member, for example, is usually a strong tie—rooted in shared experiences, a lot of trust, a deep sense of reciprocity, and a great deal of interaction. Because birds of a feather flock together (a phenomenon network researchers call *homophily*), you're likely to have more in common with your strong ties, especially those you've chosen rather than inherited. These are your people, your proverbial tribe. A weak tie, on the other hand, is someone you might know only distantly—a business acquaintance, or someone you know distantly through a strong tie, or a friend of a friend.

In a classic 1973 study, Stanford sociologist Mark Granovetter interviewed dozens of people to find out what role their social networks had played in their landing a new job. He discovered that most had found their jobs through a weak tie—not a close friend, but an acquaintance. Ever since, the power of weak ties has been lauded in sociology and network theory; weak ties have been found to be important for all different kinds of social mobility and innovation diffusion. "Because weak ties connect people across very different neighborhoods of a social network, they are critical for quickly finding new information—like where a new job opening might be—that was unavailable to you and your immediate connections," says Sinan Aral, an information economist at New York University. These weak connections form bridges over holes in the social fabric, where people are less well connected. "This can be critical, particularly in the case of an emergency, for propagating information across a network."

But strong ties have a vital place too. "Having a team built on strong ties—close connections—is essential for the intense, collaborative work of creating or synthesizing new material, especially complex material," says Aral.

In fact, Aral's research suggests that Granovetter's insight about the strength of weak ties may be incomplete. For all of the connectivity that weak ties provide, most people still get the majority of their novel information from their strong ties, not their weak ones. That's because the frequency and intensity of the interactions between you and your tribe overwhelms the relatively infrequent and narrower exchanges you have with your acquaintances. "Let's say you talk to your best friend three times a week and an old golfing buddy once a year. Even if only a tiny portion of what your best friend tells you is new, in the end, that's still, proportionally, where most of your new information is going to be coming from," says Aral.

This is part of a larger phenomenon called the *diversity/bandwidth trade-off*. As we expand the diversity of social connections we have, the bandwidth we can commit to each of those connections becomes more limited, and the information that comes across them gets weaker and narrower. And that in turn makes weak ties suitable for certain kinds of work and strong ties suitable for others.

"The most powerful constellation seems to be to have small, diverse teams of strongly tied collaborators, who each have a large and diverse weak-tie network—the best of both worlds," adds Aral.

We see the interplay of strong and weak ties throughout the story of 4636. A platform like Twitter, for example, makes it incredibly easy to maintain weak ties—to passively follow people you don't know well, and through them, to get to people you might need in an emergency. And that's exactly how Josh Nesbit used it to find Jean-Marc Castera at Digicel. That was a linchpin event—one of many—and had it not happened, the project could not have succeeded as it did.

On the other hand, the volunteers weren't entirely random groups of strangers meeting for the first time. At the center of the project were small teams made up of people who knew and trusted one another already—small teams dominated by strong ties, like the tribe of engineers building the software platform itself.

The real power of the 4636 project came in the way it enabled weakly tied volunteers to rapidly turn into strongly tied, committed collaborators. Though only a small percentage of the collaborators had met in person, the informal network provided a baseline of modest trust. Just as William Ury hopes to tap into the narrative of Abraham to remind the three religious tribes of their shared culture and values, the open nature of the 4636 platform—both the code that powers Ushahidi and the collaborative nature of the mapping—enlarged the tribe of disaster relief responders from officials at international bureaucratic organizations to a swarm of volunteers all over the world. The 4636 project was able to spread—as evidenced by Meier's description of the growth of the volunteer network—because the platform allowed people to be recruited in to perform discrete, useful tasks, with a minimum of formal authority.

For Ushahidi-Haiti @ Tufts, this was greatly aided by the culture of graduate school itself—a collaborative environment filled with people with lots of unstructured time who could make face-to-face commitments with that time fairly easily.

The 4636 project also had leaders who modeled desired behavior in the network—commitment to a shared goal above all else—and didn't show any sign of defecting from the common purpose. At the highest-stakes moments, players like Meier and Munro never gave up. They forged ahead with difficult tasks and offered encouragement for others to do so as well. The InSTEDD team worked tirelessly, and often thanklessly, behind the scenes for the good of all. Josh Nesbit set his own self-interest aside by taking his technology out of the mix, instilling trust by presenting himself as a neutral arbiter and connector.

Finally, there was a sense that the platform was showing results—feedback loops that motivated the team members and increased their sense of agency. The more they did, the more valuable it became; the more valuable it became, the more they wanted to do. By the time a volunteer like Sewell was on her third or fourth day of mapping, she felt not just needed but essential to the relief effort.

Yet we must be careful not to overglamorize Mission 4636. While it mostly succeeded spectacularly as a collaborative effort (which is why we explore it here), there are also some important places where it failed.

First, because the entire system was being built in real time, there was no process for deep integration with systems that drive traditional large-scale disaster response. Responders arriving in Port au Prince in the days following the quake faced a near-complete information gap: Many of the most basic datasets that describe a country's infrastructure, its roads, hospital networks, schools, and water systems, were on computers now buried in the rubble. Many of the people who "owned" that data, whether in Haitian government, the UN, or various NGOs, were missing or dead. The country's Ministry of Education, for example, was obliterated, and not a *single* list of schools—where one might house the homeless—survived. Finding and reconstructing these datasets obsessed the formal response organizations in the first hours after the quake. In this sense, there were important mismatches between what the formal institutions felt they needed, what the crisis mappers were providing, and how they were providing it. Paradoxically, even as they were missing the "usual" data on which to proceed on the ground, the volume and format of 4636 data gave some of the traditional players a sense of information overload—they were neither prepared for it, nor had protocols for how to act on it.

There was also a question about how crisis-mapping information was authenticated. How accurate were the reports? People in a crisis have strong incentives to sound alarm bells—with limited resources, how could the mappers and responders validate that the needs were real? This was why, on the ground, the Ushahidi maps were frequently used less to respond to singular events and more to establish emergent "centers of gravity"—trends and clusters in the data that showed where needs were emerging or still unmet.

Finally, there was a simple matter of scale. The new and nimble technology players were mostly tiny, and a few were barely organizations at all. Few had the infrastructure and proper resources needed to deliver replicable results—some could barely cover the cost of pizza to feed their volunteers. Weeks after the Haitian earthquake, an even larger quake hit Pakistan. Yet nothing like the collaborative 4636 model emerged. Why? Partly it was simple fatigue, and the continuing deep involvement of many of the best players in Haiti. The remoteness of the location from North America was also a factor. And there was a deeper issue: Haiti had represented a significant opportunity for many of these new, smaller organizations to demonstrate their value to their funders. Yet the process of sorting out who had contributed what, and how important each element was, was messy. Some organizations' roles were overemphasized, and others underemphasized, leading to some hurt feelings behind the scenes. Especially in such fluid, collaborative contexts, proper credit and proper rewards matter all the more.

Yet for all of these concerns, there can be no doubt that the 4636 effort was a success. For the first time, a global community of technical volunteers, affected citizens and global diaspora communities made material contributions to a major disaster response. Many of the failures were unavoidable by-products of the effort's novelty and will be addressed with improvements and the development of new protocols for next time.

The lessons are not just for disaster relief. Think back one last time to our bankers around the table at 33 Liberty Street. In a high-stakes weekend humming with the undercurrents of "us" against "them," what might have happened if the structural elements in place had insured space for such a 4636-style, inclusive, collaborative, and innovative "third side" response? As we'll see in the next chapter, recruiting a more diverse array of players—enlarging the tribe through both technological platforms and collaborative processes—not only might have changed the moral vector of the discussions, it could have opened up entirely new vistas in the modes of thinking around the table.

6

COGNITIVE DIVERSITY

The first gasoline-powered horseless carriages started to appear on English roadways in the early twentieth century, and with them came an entirely new domain for civil management: traffic and road safety. Not only were there no street signs, or even clear rules of the road, the roads themselves were not designed with motorists in mind. And so a social movement was launched to improve them. For its part, the Motor Union of Great Britain and Ireland suggested that the owners of British estates clip their high hedges to enable drivers on the roads to see over them.

In response, on July 13, 1908, the following letter appeared in the *Times* (London), dashed off in haste by an angry gentleman named Colonel Willoughby Verner.

Dear Sir,

Before any of your readers may be induced to cut their hedges as suggested by the secretary of the Motor Union they may like to know my experience of having done so. Four years ago I cut down the hedges and shrubs to a height of 4ft for 30 yards back from the dangerous crossing in this hamlet. The results were twofold: the following summer my garden was smothered with dust caused by fast-driven cars, and the average pace of the passing cars was considerably increased. This was bad enough, but when the culprits secured by the police pleaded that "it was perfectly safe to go fast" because "they could see well at the corner," I realized that I had made a mistake.

Since then I have let my hedges and shrubs grow, and by planting roses and hops have raised a screen 8ft to 10ft high, by which means the garden is sheltered to some degree from the dust and the speed of many passing cars sensibly diminished. For it is perfectly plain that there are a large number of motorists who can only be induced to go at a reasonable speed at cross-roads by consideration for their own personal safety.

Hence the advantage to the public of automatically fostering this spirit as I am now doing. To cut hedges is a direct encouragement to reckless driving.

Your obedient servant,
Willoughby Verner

Are all safety benefits simply consumed as performance benefits? Are human beings fundamentally wired to try to game the system for greater efficiency and reward? Engineering efforts are forever trying to help us become safer as a society but, ironically, these same innovations often push us closer and closer to the threshold of danger.

In 1815, Sir Humphry Davy, president of the Royal Society, invented a safety lamp for miners—the Davy lamp—that earned a reputation as one of the most significant safety improvements in the history of mining. When the lamp was put into use in the country's mines, however, not only did the explosions and fatalities not decrease in number, they actually increased. How could this be?

As it turned out, the lamp operated at a temperature below the ignition point of methane, thereby permitting the extension of mining into methane-rich—and increasingly dangerous—atmospheres. What started out as an attempt at a safety measure ended up only pushing the system closer to its edge. This tendency—to slowly erode the controlling mechanisms that are present to ensure safety—is a central dynamic in many robust-yet-fragile systems, especially social ones.

In 1975, Sam Peltzman, a University of Chicago economist, analyzed all the federal auto-safety standards imposed throughout the late 1960s. In his published account, he concluded that though these standards provided greater safety for vehicle occupants, they also led to the deaths of pedestrians, cyclists, and other drivers on the road.

John Adams, professor of geography at University College London, has written on the subject of risk for more than two decades. In 1981, he published a now famous study on the impact of seat belts on highway fatalities.

"Why, in country after country that mandated seat belts, was it impossible to see the promised reduction in road accident fatalities?" Adams wrote in one of his many essays on risk. "It appears that measures that protect drivers from the consequences of bad driving encourage bad driving. The principal effect of seat belt legislation has been a shift in the burden of risk from those already best protected in cars, to the most vulnerable, pedestrians and cyclists, outside cars."

Adams, along with a growing cadre of behavioral scientists and risk analysts, started to group these counterintuitive findings under the concept of *risk compensation*, the idea that humans have an inborn tolerance for risk. As safety features are added to vehicles and roads, drivers feel less vulnerable and tend to take more chances. The feeling of greater security tempts us to be more reckless.

The phenomenon can be observed in all aspects of our daily lives. Children who wear protective gear during their games have a tendency to take more physical risks. Public health officials have noted that improved HIV treatment has led to riskier and riskier sexual behavior. Forest rangers report that hikers take more risks when they think a rescuer can access them easily. Perhaps no one has put it better than Bill Booth, famous skydiver, who coined Booth's rule number 2: "The safer skydiving gear becomes, the more chances skydivers will take, in order to keep the fatality rate constant."

Most social scientists agree that risk compensation exists but, in 1976, Gerald Wilde devised a model that pushed the theory of risk thresholds even further. In collaboration with John Adams, Wilde postulated that humans are constantly balancing risk, safety, and reward in a dance that is analogous to a thermostat. The setting of the thermostat varies from one individual to another, from one group to another, from one culture to another. According to Adams, "Some like it hot—a Hell's Angel or a Grand Prix racing driver, for example; others like it cool—a Mr. Milquetoast or a little old lady named Prudence. But no one wants absolute zero." Wilde called this the "theory of risk homeostasis": All individuals become accustomed to some acceptable level of risk—a risk temperature—so when they are required to reduce risk in one area of their life, they will find themselves, consciously or unconsciously, increasing other risks until they are back in their risk temperature comfort zone. If they are required to wear seat belts, the evidence suggests they drive faster, pass other cars more dangerously, put on makeup while driving—you name it—just to stay in their comfort zone. In effect, they consume the additional safety they are required to have by changing their driving behavior so as to attain other desirable ends.

The concept of homeostasis can be seen at work in many biological systems. Despite the widely fluctuating temperature outside, our bodies maintain a core body temperature around 98.6 degrees Fahrenheit. If the heat outside notches that number up, count on our bodies to perspire to cool us down. Such regulating mechanisms make up the essential machinery of all living systems. When the brain doesn't have enough glucose to function properly, the liver kicks into gear and releases glycogen, transferred by the blood to restore levels in the brain, maintaining homeostasis in the whole system.

There is even evidence that organisms exhibit homeostasis at the population level. Mice kept in cages, amply fed and without any predators, do not grow in numbers indefinitely. Scientists find that after a certain optimum population is achieved, there is a decrease in ovulation and reproduction in the females. Guppies kept in tanks exhibit the same population homeostasis, but their numbers are kept in check by cannibalism of the young just after birth.

Risk homeostasis functions through similar feedback mechanisms. Think of your own household furnace. This is a simple but elegant example of homeostasis. The thermostat is connected to a thermometer by a switch so that when the temperature in the house drops to the level selected, that same switch turns on the furnace. As the furnace brings more and more heat into the house, the thermometer rises until it reaches the level of homeostasis—the optimum temperature—and then it switches off the furnace and shuts down the heat. When the temperature cools down again, that information is fed back into the furnace through the thermostat and the whole cycle begins again.

These same oscillations, Wilde argues, occur when our risk thermostats are in a feedback loop with our surroundings, leading us to modify our behavior. Most of us slow down when we are driving in a snowstorm. Some men and most women think twice before walking down a dark alley in an unknown part of town. If a floor is slippery, we are more likely to tread carefully. These behaviors will strike us as simple common sense.

Less intuitive is the risky behavior that creates a balancing feedback loop to the safety mechanisms of policy and regulation. A three-year study of taxicabs in Munich, Germany, concluded that cabs with antilock brakes had slightly more accidents than those without. Even more disturbing, the cabs' accelerometers showed clearly that the drivers with the safer cars accelerated faster and stopped harder.

Perhaps the most insidious example is the oft-cited study by Kip Viscusi of Duke University regarding the Consumer Product Safety Commission's mandate for childproof aspirin caps:

> A much more surprising result was the pattern displayed by poisoning rates after the advent of safety caps. For those products covered by safety caps, there was no downward shift in poisoning rates. This ineffectiveness appears to be attributable in part to increased parental irresponsibility, such as leaving the caps off bottles. This lulling effect in turn led to a higher level of poisonings for related products not protected by the caps.

Wilde likens our appetite for risk to a river delta. If the river divides into three channels before emptying out into the ocean, we can't simply dam up two of the three channels and make the flow of water disappear. Our desire for risk, like a flowing river, will widen out the one remaining channel or open entirely new channels. In other words, if we make a risky activity like skydiving illegal, Wilde posits, all of the former skydivers are not likely to take up basket weaving. Instead they will innovate and develop new risk-laden fads until they are back in their risk thermostat comfort zone.

What if Wilde's theory applies not just to individual choices but, more disconcertingly, to the culture of an entire community or organization?

CULTURES OF RISK

By now, the legacy of the BP Deepwater Horizon oil well disaster on April 20, 2010, which killed eleven and caused the largest man-made environmental catastrophe in U.S. history, is well documented.

Less appreciated, at least at the time, was that this disaster was part of a regular pattern with BP. Throughout the prior decade leading up to the spill, a serious catastrophe had been associated with the company about every other year. In 2003, a BP rig in the North Sea experienced a massive upwelling of gas that nearly destroyed the platform; in 2005, a BP refinery exploded in Texas City, Texas, killing fifteen workers; in 2006, a BP pipeline on the North Slope of Alaska ruptured, spilling 200,000 gallons of crude oil.

In 2007, Carolyn Merritt, chairman of the U.S. Chemical Safety Board, led an investigation of the Texas City explosion and noted, "As the investigation unfolded, we were absolutely terrified that such a [poor safety] culture could exist at BP."

That same year, following a sex scandal, BP replaced its larger-than-life CEO, Lord John Browne. Browne had led the company for a decade, but he had earned a reputation for neglecting day-to-day concerns—including safety—at the expense of large-scale deal making. Incoming CEO Tony Hayward noted that BP's practices "failed to meet our own standards and the requirements of the law" and promised to "focus like a laser" on the company's accident record.

Unfortunately, BP's culture proved resistant to Hayward's beam. In 2009, two years into his term, the Occupational Safety and Health Administration (OSHA) documented more than seven hundred violations at the very same Texas City refinery where the deadly 2005 explosion had occurred just four years before. The agency fined the company $87.4 million, more than four times the amount it had fined the company for the original 2005 explosion. Dangerous situations also continued to plague the company's operations in Alaska during the same time period.

These fines and admonitions had little impact on the company, whose leadership increasingly came to see them as the cost of growth in the high-stakes world of global energy extraction. Instead, BP baked high-risk decision making into its engineering and operations plans, replacing normal procedures with ones designed to save money.

On the day of the Deepwater Horizon accident, for example, BP decided to replace heavy drilling mud with lighter seawater to seal the well; the maneuver was designed to accelerate a process that was running behind schedule and costing the company some $750,000 a day. But the procedure was untested, and workers on the rig, including the chief driller, Dewey Revette, expressed grave concerns. They were overruled. Just as the workers feared, the lighter material provided insufficient downward pressure to keep errant gas from escaping, leading directly to the blowout. Revette and ten others were killed.

According to survivors of the disaster and other former employees of BP, for years it was tacitly understood that if you raised safety concerns, you could get fired from the company. Oberon Houston, a former employee who had previously served as the deputy offshore installation manager of a BP-operated production platform in the North Sea, commented in a recent blog post, "BP management focused heavily on the easy part of safety, holding the hand rails, spending hours discussing the merits of reverse parking and the dangers of not having a lid on a coffee cup, but were less enthusiastic about the hard stuff, investing in and maintaining their complex facilities.

"A continual focus on costs and an undoubted commercial savvy was not complemented with similar expertise, or enthusiasm, for the nuts and bolts of the job," Oberon added. "Management listened intently to the views of market analysts, who knew little about the technical detail of the oil business, but instead were driven by quarterly results; encouraging and cheering on management's relentless drive to reduce costs. This resulted in a chronic short term view at the very top of the company."

The Deepwater Horizon spill epitomizes the central role of culture in both amplifying a mediating risk and creating or obliterating the conditions for greater organizational resilience—and not just in the world of offshore drilling. For example, in a 2009 survey conducted of almost five hundred bank executives by the consulting firm KPMG,

almost half—48 percent—of respondents cited the financial firms' risk culture as a leading contributor to the financial crisis. More than half—58 percent—of corporate board members and internal auditors included in the survey said that their company's employees had little or no understanding of how risks should even be assessed.

In the absence of any contravening internal signals, one or another level of risk homeostasis naturally takes root, and increasingly narrow styles of thought take hold. Those who espouse the dominant perspective and values are rewarded and promoted; those who espouse different norms can be systematically undermined and driven away. For good or ill, every time that happens, silent but powerful cues are sent to every individual who remains: *You can't change it. It's just how things are done here. This is how you need to think and act if you want to succeed.* Think back to the way Geoffrey West championed the dynamism of cities through their diversity, the presence of "crazy" people. Without these crazies—people capable of dissent—fragility sets in.

Can anything be done to reverse the resulting myopia? As the U.S. Army is discovering, one way is to build and deploy a corps of professional skeptics.

RED TEAM UNIVERSITY

Fort Leavenworth, located in its namesake town of Leavenworth, Kansas, is the oldest active army post west of the Mississippi River, in continuous operation since 1827. Given its heritage and stately architecture, the casual observer might confuse many of its buildings for those of a staid eastern college campus, yet it's actually home to a far more visionary educational institution: the University of Foreign Military and Cultural Studies, more commonly known by its nickname, Red Team University.

Started in 2004 by a group led by retired army colonel Greg Fontenot, Red Team U is an effort to train professional military devil's

advocates—field operatives who bring critical thinking to the battle-field and help commanding officers avoid the perils of overconfidence, strategic brittleness, and groupthink.

As expounded upon by Irving Janis in 1972 to describe fiascos like the Bay of Pigs invasion, groupthink is an organizational pathology that can occur within any tightly knit group of people who depend on social cohesion to operate—a nearly perfect working description of units of fighting soldiers on the battlefield. Its hallmarks include a strong illusion of invulnerability by key decision makers; a belief in the inherent morality of the group; the stereotyping of those who do not agree with the group's perspective; and overly simplistic moral formulations that dissuade deeper rational analysis. Self-appointed thought-guards prevent alternative views from being aired and place significant pressure on dissenters, leading to the illusion of unanimity, even if dissent is rampant below the surface. It's this kind of cultural and cognitive insularity that can get soldiers killed and needlessly prolong wars, and Fontenot and his team are on a mission to stamp it out.

Fontenot is bald with dark glasses, a penchant for cigars, and a piercing, no-nonsense gaze that suggests he may have recently made more important decisions than speaking to you. He appears to be every bit the straight-from-central-casting, front-line tank commander he was for almost three decades. Until he opens his mouth, at which point he may, in a measured drawl, turn the conversation with equal erudition from an analysis of military strategies against asymmetric threats like al-Qaeda, to the Jungian archetypes of North Korean leaders, to culturally illuminating concepts of Chinese philosophy. This is, after all, a man who taught history at West Point.

Fontenot's extensive military career had given him a front-row seat to the entire parade of post–cold war battles fought by the United States and firsthand experience with both the evolving purposes to which American military power would be seconded and with the creeping dangers of groupthink. He commanded a battalion in the army's point division that broke the Saddam line in the First Gulf War

and later commanded the first brigade to enter Bosnia. But he was unprepared for what he found when he arrived.

"I had studied World War I extensively, and so I was excited to get to Bosnia. But when I got there, I found we didn't understand the operating environment at all. Everyone was killing one another over issues only they understood. To us, it was incomprehensible. Everyone looked the same, and we couldn't imagine how they perceived us. It was humbling. Walking in, I thought I knew something about it. I quickly realized I knew nothing."

With insufficient cultural awareness, Fontenot watched as units on the ground struggled to make sense of it all. When in doubt, the soldiers frequently reverted to what they knew, falling back into rote practices and mind-sets that were not necessarily well adapted to the circumstances in which they found themselves. Echoing Arquilla's predictions of wars yet to come, the Department of Killing People and Breaking Things were being asked to try to put things back together, or at least keep things from falling apart further.

"In that kind of situation, culture is everything," says Fontenot. "It's not just about understanding the combatants' capabilities, it's about understanding how they think. The Balkans have historically been a bad neighborhood, with ancient enmities that were largely opaque to outsiders. Our forces were there with the right hardware, but without the right cultural software, the likelihood of things being misinterpreted goes up dramatically. One man's act of deterrence is another's act of war."

To combat surprises in the field, militaries around the world have long engaged in war gaming. In these simulated rehearsals of real conflict, the home team is assigned the color blue and the enemy the color red. The blue team develops plans for the exercise, while the opposing red team attempts to either defend a position or disrupt blue's operations. There are significant limitations to these exercises: For one thing, such games commonly hew to the blue team's plans, without allowing the red team to alter its strategy and influence the blue team's tactics. But even so, war games enable militaries to think about how enemies, civilians, and partners—collectively "the others" on a battlefield—might respond to a hypothetical situation. In an age dominated by increasingly complex, low-intensity, coalition-style fighting and outpost and outreach antiterrorism and counterinsurgency operations, the need for such insights has accelerated dramatically.

Fontenot wanted to take this "think like the others" war gaming approach to new heights of sophistication, remove it from the simulated context altogether, and embed it in real units doing real fighting and real peacekeeping. In 2004, Red Team University was launched; today almost three hundred graduates are operating in the field around the world.

The eighteen-week Red Team course offers an intensive, and intentionally eclectic, survey of connected ideas in everything from military theory to strategies for negotiation, business modeling to terrorism and counterinsurgency, mixed with case studies and a heavy dose of anthropology.

Fontenot and the instructors are constantly evolving the syllabus, which relies on a mix of well-known works on creative thinking and behavioral economics; strategic in-depth analyses of current concerns like Iraq and Iran, the Middle East, terror networks, and North Korea; and lesser-known treatises on philosophy and cultural criticism. A good example text is *The Propensity of Things* by French sinologist François Jullien, which explores the Chinese notion of *shi*, a term with multiple connotations and no direct English translation, which is intrinsic to a wide array of Chinese thought, military and otherwise. *Shi* encompasses notions of power, relationship, and circumstance, though Jullien translates it as "propensity," or a tendency that, like a seed, germinates within a situation. Once the propensity of a situation is set off, it can't be stopped until the situation comes back into equilibrium. Thus, according to Chinese thought, a great power imbalance contains within itself not merely the potential, but the propensity for a great rebalancing. If one understands and designs for the propensity of the actors on a battlefield and can shape the energetic forces that are already playing out, conflict itself may be avoided, even as the desired outcome is achieved.

Like many of the concepts introduced at Red Team U, learning about *shi* requires embracing new cultural frameworks and new modes of thought, in this case a shift from thinking about stockpiles and objects—the standard units of military force—to flows and relationships. This is a style of thinking to which few of the participating officers have previously been exposed.

The broad conceptual portfolio is a welcome respite for many uniformed officers who are used to narrower, more top-down teaching approaches, but this is not some lightweight seminar: There are 250 pages of reading and analysis, on average, every evening. The program is designed to encourage Red Team officers to think laterally, challenge and de-bias their assumptions, ask tough questions of a commander in the field, and help them consider less-obvious cultural perceptions of and by U.S. coalition partners, adversaries, and others. "The goal," says Fontenot, "is to help them to escape the conventions of the Western military mind-set, so they can in turn help others see past it."

Of course, making connections in a classroom is one thing; working with commanders in the field is another. Initial skepticism about the Red Team University approach in the field is expected; graduates sometimes face outright hostility. "Some of the flak results from the nature of the command-and-control structure itself, some of it is just a by-product of people working extremely hard and not being able to step back and see the larger picture," says Steve Hall, a graduate of the program. "The senior guys, they want this, but some of the tactical guys give us some pushback—they say, 'I don't need someone looking over my shoulder.'"

That's why a big part of Red Team U's curriculum is focused on how to effectively pitch new ideas to soldiers in the field. "Our job isn't to second-guess the commanders, it's to help them become better

thinkers—to consider perspectives and options that they wouldn't normally," adds Hall. Like all Red Team University graduates, he was
taught to raise issues, then back away if things become too conten-
tious. Pushing too hard can paralyze a group with indecision—exactly
the opposite of what Red Teamers trying to achieve. "We focus on the
psychology of positive reinforcement, of suggestion as much as respectful challenge. You have to figure out how to sell the boss." During the
program, officers deconstruct scenes from films like *The Godfather*, exploring the tactics of the consigliere character played by Robert Duvall
to mine ideas for effectively communicating externalities up the chain
of command.

Hall adds that Red Team University's approach is particularly
timely, given that much of the training received by today's officers was
designed for entirely different kinds of conflict. "I was trained as a
Cobra helicopter pilot," says Hall. "We were trained to think in terms
of dealing with the Soviet Union. We would fly our helicopters low
and slow, at night, to avoid radar—and we'd worry about tanks. Now, in
Iraq, none of that is relevant: You fly during the day, there is no radar
to avoid, and you worry about Kalashnikov machine guns, an outdated
weapon that can still kill you."

"Our underlying assumption is not that people are evil, lazy, or incapable, but that it's just hard to critique your own work when you're
doing it," adds Fontenot. "People reason by analogy, and it's hard to
recognize your own untested hypotheses. If someone doesn't challenge
them, hubris can set in, bred of custom and complacency."

In breaking up that complacency, Red Team increases what Scott Page,
a professor of complex systems, political science, and economics at the
University of Michigan, calls the "cognitive diversity" of the team—the
distribution of different kinds of thinkers within each group. Mathematical modeling suggests that improving cognitive diversity within
a team can lead to vastly better outcomes. Page points to the diversity
prediction theorem, an empirically tested mathematical model that

shows that a crowd's collective accuracy equals its average individual accuracy minus its collective predictive diversity.

What does that mean exactly? To achieve a truly wise crowd with accurate predictive skills, you either need to have an extremely smart crowd (high ability) or you need to have people who are moderately smart but who also happen to be cognitively diverse (high diversity). Both ability and diversity contribute equally and positively—a deeply diverse team can be as good as a deeply talented one. Using a method similar to the portfolio approach we discussed in chapter 1, remixing talent into highly cognitively diverse teams produces safer, more resilient, and better performance.

The performance benefits of cognitive diversity have been experimentally validated by Kevin Dunbar, a psychologist at the University of Toronto who has explored its impact in another highly rigorous and regimented setting: scientific research labs. Over the course of a year, Dunbar and his team studied the working habits of four molecular biology labs. In a turnabout of normal affairs, the scientists were the subjects, and Dunbar studied them in their habitat, much as a primatologist studies chimps in the wild. He attended regular lab meetings where scientists present their research and current problems to one another; reviewed their data and their interim work product; and spent time in the lab interviewing, observing, and just hanging out.

What did he find? Unlike stereotypical notions of science as a rational, if tedious process, "the actual process of science is surprisingly sloppy and fraught with uncertainty—you are constantly confronted with outcomes you didn't expect," says Dunbar. "When scientists gets a surprising result—which can be as often as fifty percent of the time— they have to ask themselves: Was this the result of some methodological error on my part, a problem with the equipment, or is it a significant new result? What does it mean?"

When answering those questions, Dunbar found that scientists, like most people, tend to explain unexpected results through analogy. But there were big differences in how groups of scientists reasoned together.

At lab meetings with lots of scientists from a single field, unexpected results tended to be interpreted through the lens of narrower, local analogies. For example, in a lab filled entirely with *E. coli* researchers, unexpected results would be interpreted almost exclusively in the light of prior *E. coli* findings, and these labs generally made slower progress. In contrast, labs filled with a more diverse array of scientists tended to use broader, more long-distance analogies, drawing on concepts and prior results well outside the target field of study, and tended to progress more quickly.

Dunbar makes (what else?) an analogy: "Local analogies are like pawns on a chess board—they only allow you to move in limited ways and require you to exhaustively check a large number of modestly different possibilities. Longer-distance analogies, on the other hand, are like queens, allowing you to rapidly travel to entirely different parts of the search space of solutions."

The effect can be profound. "By coincidence, in one week, we happened to observe two different molecular biology labs confronting exactly the same technical problem. Lab A was headed by a brilliant scientist but a rather self-similar scientific staff. Lab B was filled with diverse scientists—a chemist, an MD, a geneticist, and so on. Two members of Lab B solved the problem in two minutes at a meeting, and Lab A was still working incrementally on the problem two months later."

The diversity of Lab B is not free—it comes at a cost: It takes additional time to harmonize and integrate diverse team members, which can be seen as imposing a multidisciplinarity tax and modest suboptimality on the group, in exchange for dramatically better results getting out of the inevitable jams when they occur. And the members can't be so diverse that they diverge in terms of values, differ on the end goal, or can't bridge their methodological differences. "We found the labs that harnessed their cognitive diversity occupied a warm zone where there were meaningful disciplinary differences between team members but not irreconcilable ones. Most important, the members were united by a commonly understood goal, whether that was hunting a virus or unlocking a gene mechanism. When all of those pieces are in place, the team members were motivated to spend lots of time explaining themselves to one another and developing their own shared terms and language; new participants end up learning that special, cross-disciplinary language as part of the process of integration."

Equally important is that these teams employed the right kinds of analogies at the right time. "Over and over again, we found that local analogies were useful for fixing experiments that weren't working; regional analogies for coming up with new hypotheses; and long-distance analogies are good for explaining things to nonspecialists."

Dunbar's warm zone findings with scientific teams are mirrored by additional research by Sinan Aral at New York University. Aral and a team of colleagues studied the email correspondence of nearly 1,400 teams of executive recruiters over five years and found that graphs of their productivity followed an inverted U shape. Like the scientists, when the recruiters were too closely aligned in terms of their subject-matter expertise and social networks, they were less productive at finding good candidates for open positions. On the other hand, when the recruiters were working far outside their area of expertise and their social networks, they were similarly less productive. But in the middle of these two extremes, where team members shared the same mental models but brought a moderate amount of diversity to the tasks at hand, productivity skyrocketed. "With the right amount of diversity, the volume, speed, and revenue generated by the work all went up."

Dunbar's team also found that cognitive diversity and actual diversity reinforce each other. For instance, as a rule, when male scientists received an unexpected finding, they assumed they knew what the cause was and went ahead anyway, which frequently sent them charging down blind alleys. Female scientists, on the other hand, tried more often to replicate results in order to find out why they got the unexpected findings. "There's a pernicious view that women are passive or that only those who act like men can compete," Dunbar said. "But in these labs we studied, the women were just as aggressive as the men—they just approached the unexpected in a completely different way."

In addition to documenting the way these scientists worked, Dunbar also studied the way their brains responded to both expected and unexpected results. In cases where subjects were confronted with an expected result, the areas of the brain responsible for putting that information into memory were activated, akin to a reward for good behavior: *I got what I wanted, I'll remember this!* When confronted with an unexpected result, however, the information was often not committed to memory at all. "Sometimes, data that you don't like, you don't even process," Dunbar said.

This is the neurological manifestation of the widely observed confirmation bias, a person's tendency to favor information that confirms his assumptions, preconceptions, or hypotheses whether they are actually true or not. It was first observed by the English psychologist Peter Wason in 1960, who demonstrated it with a simple but powerful test. Wason would present experimental subjects with a triplet of numbers, say 2–4–6, and each would have to guess what rule tied the three numbers together. Subjects would do this by providing their own triplet back to the experimenter, who would either confirm the guess as conforming to the rule, or not.

When presented with the triplet above as a starting point, most subjects started by forming an initial hypothesis about them: They are a sequence of even numbers. They would test this hypothesis with a few guesses, such as 4–6–8, 4–8–12, and then perhaps 8–10–12, all of which would generate a positive confirmation from the experimenter, and then they would stop, confident they had confirmed the rule. Only they hadn't: The rule was that the triplet contained ascending numbers, not even numbers. Surprisingly few people generated any guesses that would disconfirm their hypothesis (such as 8–6–4, 1–2–3, or 2–2–2) and only one in five subjects guessed the underlying rule correctly.

The subconscious aversion underlying confirmation bias is rooted in our universal distaste for finding out that we might be wrong. For evidence of its power, try a simple experiment: Say something considered politically opposite of your beliefs in a forum like Twitter or on Facebook, and almost certainly you will see some instantaneous churn in those who follow you, as some stop and others start, each manifesting his or her own confirmation biases.

This phenomenon not only causes us to avoid messages we don't agree with, it can shade our interpretation of those we can't avoid. In 2009, Heather LaMarre and her colleagues at Ohio State University found that when watching *The Colbert Report*, politically conservative U.S. viewers were more likely to report that Colbert disliked liberalism, only pretends to be joking, and genuinely meant what he said, while liberal viewers were more likely to report that Colbert used satire to mock conservatives and was not serious when offering conservative political statements with which they disagreed. While everyone agrees he's funny, for the average viewer, the interpretation of whether Colbert is actually a liberal or a conservative is strongly predictive of whether the viewer is actually a liberal or conservative: The show serves as an elegant mirror, affirming our identity even as it confirms our prejudices.

If we so readily make mistakes about something as ubiquitous and designed for relevance as *The Colbert Report*, imagine the challenge for the army's Red Teamers, embedded with forces operating in culturally foreign and sporadically lethal environments, where a command-and-control organizational structure dominates, and where confirmation bias can literally kill.

To be effective, they must challenge—supportively—where they can, expanding the conceptual search space of senior commanders' thinking. They must give voice to unpopular, unconventional, or unorthodox views of strategies that may have been authored by the very commanders they serve. They must maintain a warm zone of respectful challenge and supportive dissent, becoming neither co-opted nor ostracized by the larger chain of command, while confronting the omnipresent dangers of homeostasis, groupthink, and confirmation bias. It can be difficult for the Red Teamers to show they're having an impact, as

theirs is a long game of institutional cultural change, of vigilantly challenging implicit norms and biases before they harden. If they're doing
their job well, they leave no fingerprints, just better decisions in their
wake.

For their commanders, and indeed all leaders, the lessons are also
clear: Resilient cultures are rooted in diversity and difference and are
tolerant of occasional dissent. These factors protect the alternative
search spaces that are so vital to any community struggling to change
maladaptive cultural norms. The success of Red Team University—an
intervention designed by the military for the military—is due, in no
small part, to the social credibility of its professional skeptics. Though
they are trained to be outsiders, they are still very much in the culture, allowing them to work in the diversity warm zone. As we'll see
next, embedding such resilience-enhancing interventions authentically
within a community's culture in this way isn't just preferable—it's essential to making them work.

7

COMMUNITIES THAT
BOUNCE BACK

Like the coral reefs and economic systems we discussed at the beginning of this book, every community has to respond to emerging
internal problems from time to time—maladaptive behaviors, issues,
or circumstances that, if left to fester, can tip it past a critical and unwanted threshold. Whether it's stopping a behavioral contagion (like
needle sharing among drug users) or challenging a social norm (like
getting people to recycle their trash), social resilience often rests on the
adaptive capacity of a community, or its ability to sense, interdict, and
intervene.

This capacity cannot simply be imposed from above—instead it
must be nurtured in the social structures and relationships that govern
people's everyday lives. Two efforts at community resilience, undertaken on opposite sides of the world—in Bangladesh and Chicago—
illuminate the terrible consequences of getting such interventions
wrong and the transformational power of getting them right.

THE POISON IN THE WELLS

In the early 1970s, Bangladesh was not a healthy place to live. Reeling from natural disasters, an exploding population, and abject poverty, the young, destitute country had nothing like the resources required to support its people. Nowhere was this seen—and felt—more severely than in access to clean water. Nearly a quarter of a million Bangladeshis died each year from diseases like cholera, typhoid, and hepatitis, the same number of people who died in all of the planet's natural disasters—including earthquakes, heat waves, floods, volcanoes, typhoons, blizzards, landslides, and droughts—in 2010 combined.

In 1972, development circles were abuzz with the promise of a new technology that held hope for the Bangladeshi crisis: the hand pump. Unlike in other countries, where drilling through hard rock was an expensive and arduous procedure, Bangladesh's alluvial plains consisted of soft layers of sand and clay. In only one day, three or four men could drill a 100-foot well and reach the pure water underground. UNICEF instigated a nationwide program to sink shallow tube wells across the country. Once a small hand pump was installed to the top of the tube, clean water rose quickly to the surface.

Before long, every village wanted one; the drilling teams could not sink wells fast enough. By 1978, more than 300,000 tube wells had been sunk with UNICEF's assistance; by the late 1990s, this number had swelled to more than 10 million, surpassing UNICEF's target of supplying 80 percent of Bangladeshis with safe drinking water by the year 2000.

The tube wells appeared to be a shining success story of modern development. With access to clean water, the mortality rate for children under five plummeted from almost 24 percent in 1970 to less than 10 percent in the late 1990s. UNICEF's approach to the challenge was touted as a model for South Asia and the world.

To accompany the new wells, the Bangladeshi government initiated public campaigns urging its citizens to switch from groundwater to the safer tube wells. The message was so effectively communicated that the tube wells became status symbols of wealth, security, and safety, often included in a new bride's dowry.

But in 1983, a young dermatologist, Dr. K. C. Saha at the School of Tropical Medicine in Calcutta, began to see something disturbing. Patients were walking into his clinic from nearby villages with dark markings on their skin that he noted as "black raindrops." Researching this hyperpigmentation in his patients, Saha finally determined that the markings were the first indication of an illness called *arsenicosis*, or long-term exposure to arsenic.

Colorless and odorless, arsenic is impossible to detect without a chemical test, earning it a starring role in the plots of mysteries, thrillers, and political assassinations. When arsenic concentrations are high, it can quickly cause skin lesions but, in less concentrated amounts— diluted in drinking water, for example—it has a long latency period that can last years. The people of Bangladesh were drinking poison in long, slow sips. They just couldn't taste it.

By 1987, Saha had seen more than 1,200 cases in his clinics. His patients had only one thing in common: They all drank from the same village tube wells. Sure enough, when he tested the water from these wells, Saha found arsenic in dangerously high concentrations. UNICEF had mistaken deep water for clean water and never tested its tube wells for this poison.

By 1993, forty thousand people in Bangladesh were showing skin lesion symptoms characteristic of arsenicosis, along with its other terrible symptoms: hardening of the skin, dark spots on the hands and feet, swollen limbs, and loss of feeling in the extremities. The lesions often became infected, leading to gangrene. But exposure to arsenic also contributes to more severe illnesses like skin, lung, and bladder cancer. The World Health Organization (WHO) later predicted that one in a hundred Bangladeshis drinking from the contaminated wells would die from an arsenic-related cancer.

Though Saha and his colleagues published paper after paper confirming that the drinking water was contaminated, both UNICEF's and the Bangladeshi government's response was glacial. A full decade later, in 1998, a UNICEF representative finally responded to the crisis by saying, "We are wedded to safe water, not tube wells, but at this time tube wells remain a good, affordable idea and our program will go on."

By that time, it was already too late—arsenic in the tube wells had escaped and found its way into the food supply. Rice irrigated with the tube wells was found to contain more than nine times the normal amount of arsenic. Rice concentrated the poison, so even if one managed to avoid drinking contaminated well water, concentrated amounts would show up in one's food.

By 2000, the government estimated that 40 to 50 percent of the estimated 10 million tube wells were contaminated with arsenic—one out of every two wells. The WHO made a simple but terrible pronouncement: "Bangladesh is facing the largest mass poisoning of a population in history."

In the face of impending catastrophe, it was clear that an intervention—fast, effective, and enormous in scale—would be needed. After spending the last thirty years successfully convincing the Bangladeshis to switch to tube wells, the government and NGOs now needed to reverse course, quickly communicating that some—though not all—of these wells were in fact poisonous. How would they effectively reach out to tens of millions of rural Bangladeshis, many of whom were illiterate and increasingly distrustful of government information?

Just telling people had its limitations: The word for poison, *bish*, referred to things that smelled bad or conjured up a sense of disgust. The well water, on the other hand, appeared clear and clean. The Bangladeshi government needed something simpler to convey danger, something that would resonate with a largely uneducated population. It agreed upon the universal colors of red and green.

With support from the World Bank, the government initiated a countrywide testing project called the Bangladesh Arsenic Mitigation Water Supply Project (BAMWSP). After performing arsenic field tests on wells all over the country, BAMWSP workers were then tasked with painting the spout of each well red if the well was contaminated, green if it was safe.

Five years and $44 million later, BAMWSP had field tested and color coded about half the 10 million wells in many of the country's 86,000 villages. In about 10 percent of these communities, more than 80 percent of the wells were found to be tainted—earning them the nickname "arsenic *para*," or arsenic 'hoods, because so many people living in them were marked by the telltale black raindrops all over their skin.

Officially, this intervention was hailed an almost instantaneous success. Once the wells were painted, researchers estimated that use of the red wells dropped to 1 percent. The government reported that most households felt sufficiently informed of the arsenic contamination. The international development community, once again, heralded the well-painting initiative as one of the most successful public health interventions in history.

And yet, underneath the hard data and official pronouncements, a far more complicated story started to unfold.

Instigating large-scale behavior change—in this case, getting people to switch from the red wells to the safer green wells—is a complex and multifaceted undertaking, informed by a whole host of cultural norms, taboos, incentives, and mores. The BAMWSP intervention successfully communicated wells' arsenic status to rural villagers, but the effort was implemented as a one-time testing and labeling intervention, with no corresponding measures to address the possible cultural or social implications of publicly revealing a community's well status. Unlike Willie Smits's reforestation project or the emergent organization of an effort like Mission 4636, this was an engineering-led, one-off approach, imposed from the outside.

In Bangladesh, water use starts and ends with women and girls. Domestic water—collection, storage, and distribution—is traditionally a woman's responsibility. They are the ones who will determine if a switch to a green well is warranted because they are the ones who fetch the water numerous times a day.

In a country with strict norms regarding the accompaniment of women in public, successful well switching has everything to do with where the safe wells are located and whether or not women and girls can access them in a way that is deemed socially appropriate. One academic reported that a green well was conveniently located to serve several contaminated well users but the women refused to draw water from it because it was directly in front of the mosque. The religious and cultural norms impeded a successful switch.

In the early 2000s, Farhana Sultana, a human geographer at the University of Minnesota, surveyed women and men in Bangladesh regarding their water resource use. Upon learning that their home-stead had a red well, half of the men interviewed were most concerned about their women and girls venturing out in public to get clean water. A third of the women were concerned that they would either have to travel farther to get water or use someone else's well.

Negotiating use of someone else's green well was an act fraught with potential conflict. In a village where more than 80 percent of the wells were painted red—an "arsenic *para*"—women complained that waiting in the lines at the safe wells took up too much time and squab-bling began to occur in the courtyard around the well. The owners of the green wells often expressed indignation and anger that so many women were coming onto their property. Some even removed the hand pump, allowing only their own family access to the water.

One man told Sultana, "Too many women in one place means too much noise and squabbling; who wants to put up with that daily in his own home?" Another woman claimed that she would rather drink arse-nic water than endure the constant bickering and insults while waiting in line for the green well.

Effective behavior change also requires a complete understanding of the consequences of that behavior. Although the Bangladesh De-mographic and Health Survey reported that four in five Bangladeshis "understood" the status of their well, the reality was more nuanced. According to Sultana, for many uneducated Bangladeshis, the very words "arsenic contamination" were too vague and abstract. Many of them reported that they thought the contamination was a flaw in the pumping mechanism or in the well itself because scientific concepts like groundwater and aquifers were difficult to grasp. Often, instead of fully engaging with the science, villagers would speak of the "*zengoo*" (black raindrop spots) as curses from God.

Even those villagers who reportedly understood that victims of arsenic contamination were not contagious expressed fears that one day, they would become afflicted. The villagers often agreed that it was best to ostracize those suffering from arsenicosis out of fear that the contamination would spread.

In 2003, reports started to come in of families and communities chipping away at the red paint on their wells. Some repainted their wells green, while others just left them colorless. Such was the power of the "contagious" stigma: They knowingly drank poisoned water to hide the fact that they were now tainted.

Husbands often divorced and abandoned their wives when they learned that they had been drinking from red wells, and young girls living within the vicinity of contaminated wells suffered from diminishing marriage prospects, if they were able to marry at all. In 2001, an arsenic attitudes survey reported that only one in twenty parents would allow their child to marry an arsenicosis patient, and an increased dowry was often demanded of a family when the girl was coming from a home with a red well. Researchers noted hearing a common expression, "*Beramma maiya anmu keno?*" (Why bring in a sick girl?).

In many regions around the country, the intervention successfully changed behavior—but it was a change for the worse. In the most affected areas, where more than 80 percent of wells registered an unsafe level of arsenic, only an estimated 4 million out of the total 9 million residents were provided with alternative safe water options by the government. Left without a viable, safe alternative to red tube wells, many women and girls returned to surface water sources like ponds and lakes, significantly more likely to be contaminated with fecal pathogens.

Still, the surface water sources might have been safe alternatives if the practices of filtering and boiling were still in place. Decades of tube well use, however, had eroded these habits from rural culture. What's more, the cost of fuel to burn a fire for boiling proved prohibitively expensive for most of the rural villagers. As a result of all these factors, researchers estimated that abandonment of shallow tube wells increased a household's risk of diarrheal disease by 20 percent.

Amid the gloomy outcomes, an occasional bright spot shone. In one village, Elmdi-Kamaldi, researchers from Columbia University met a self-appointed local arsenic activist, a schoolteacher with a 700-foot deep clean well. He encouraged others to drink from his well and he led the call for installation of other deep safe wells in the community. His advocacy seemed to pay off: Three in four of the owners of unsafe wells had switched to alternate wells, the highest rate of switching recorded.

What the BAMWSP initiative measured was outputs, not outcomes—the number of wells tested, not the number of people getting access to clean water or the other complex social impacts. Worse, this was a one-time intervention in a complex social system that was constantly changing.

Without deep community engagement, these problems are unlikely to turn around anytime soon. A water quality survey in 2009 by the Bangladesh Bureau of Statistics and UNICEF found that approximately 20 million people were still being exposed to excessive quantities of arsenic. In 2009, the government of Bangladesh committed in its election manifesto that the "arsenic problem would be tackled and measures taken to ensure safe drinking water for all by 2011."

And so, while the experts and politicians discuss how to find a solution for the unintended consequences of the intervention, the people of Bangladesh continue bringing their buckets to the wells while crossing their fingers behind their backs.

• • •

The story of the Bangladeshi wells is one of nested failures: the failure of government agencies to meet basic engineering competencies; the failure of well-meaning Western aid organizations to understand the environment within which they were operating; the deadly slowness of top-down bureaucracies and their inability to admit mistakes; the failure of a social intervention that took no account of the complex culture of the community intended to adopt it, or made any effort to involve the people most affected by it in its design; and the classist stance of organizations that assumed that poor people have little to contribute.

Now imagine what an energized and involved intervention might have looked like. Imagine women and girls consulted on the design of culturally appropriate solutions. Imagine a network of community agents, deputized and trained to support the well switch. Imagine local people, both embedded within the community and linked across it, monitoring, sensing, and addressing issues as they arrive, intervening where appropriate, and heading off problems before their consequences become deadly.

What you will have imagined looks a great deal like one remarkably successful intervention half a world away, on the streets of South Chicago.

ENDING AN EPIDEMIC OF VIOLENCE

Karon Clark keeps all her open cases filed in the top drawer of her thick steel desk. As an outreach worker with CeaseFire, a groundbreaking violence prevention program headquartered in urban Chicago, she carries anywhere between ten to fourteen open cases at any given time—each one a member of the community who is at elevated risk for committing violence. Some of her clients have drifted in and out of her files several times—caught in the vicious cycle between prison, rehabilitation, drugs, gangs, violence, and then prison again—but one client in particular causes her to hold her breath in hope.

"I just can't wait to shut the file on Davion. We're almost there. Davion—he's a boy who might really make it out."

Davion (his name has been changed) has been Karon Clark's client since January 2008 and, by all accounts, he has the makings of a success story. Athletic and highly skilled in football, he managed to make it through high school without ever getting mired in the drinking and drug culture. He pinned all his hopes on getting recruited for a major university football team. Then, in the final round of tryouts, he was cut.

"First day as my client, he comes in to speak to me and he's angry. 'I'm not going to college,' he says. But he's from a good family. They're not well educated but they're a good family and I told him that he still had a shot."

Clark has spent the last four months taking Davion to junior colleges around the state, introducing him to football coaches, and helping him navigate financial aid application forms. One day a few weeks ago, Clark took Davion to a junior college about an hour away for an interview.

"He met a cute girl in the library and she gave him her number. He seemed a lot more enthusiastic about college after that."

Clark takes a deep breath as she contemplates the possibility: Davion might actually be headed off to college. She has every reason to believe that she'll be removing his case file from her top drawer and putting it away in storage next year.

And yet, she's still nervous.

Clark's CeaseFire branch in the West Garfield Park neighborhood is located in an old storefront. Unlike other community organizations, CeaseFire stays open late for its clients—very late. The outreach workers keep their storefront door open until the wee hours of the morning, and every few hours they perform what they refer to as "the rounds," the daily walk around the neighborhood to check in on their clients and get a feel for the happenings out in the street.

It's a Friday night. Under normal circumstances, Clark and her fellow outreach workers would be out pounding the pavement, stopping to say hello, putting some eyes on the street. Tonight, a steady rain falls, so the group climbs into a car and starts a slow crawl through the darkened streets. At first, the corners are relatively empty. A group of young men and woman are sitting on a stoop chatting. Clark's car pulls up and they shout out, "Thank you CeaseFire! We love the T-shirts!"

"We gave them all T-shirts at our latest event," Clark explains.

One of Clark's fellow outreach workers spots his client wandering through the street: a young man in a voluminous T-shirt and baggy pants. He rolls down the car window and calls out.

"How're you doing, man?"

"Awright."

"You come and see me this week. We need to check in."

The rounds involve many of these types of encounters. Every evening, the outreach workers track down as many of their clients as they can find, gauging the daily progress of their group, offering each the resources and support needed to promote a life in mainstream society. The rounds offer CeaseFire a chance to take the temperature of the community. How hot is the neighborhood tonight?

Slowly, as the Friday night parties begin kicking into gear, more and more people come out on the street. This evening there are small groups gathering on several of the main corners. People are pulling up in dark-windowed cars with booming bass stereos. Some of the girls are teetering in high heels back and forth from car to car.

"The women cause some of the biggest fights around here," Clark comments. "The men do the shooting, but the women instigate. They provoke. They sleep with their boyfriend's friend, say. They're caught up in the same cycle of violence."

The outreach workers continue driving through the rain, slowly pointing out gathering groups of current and former clients. They stop every block or so to check in.

"What's new here?" they ask a group of teenagers starting to gather in an abandoned lot.

"Nothin'. It's cool."

Is it cool? Is it hot? The CeaseFire outreach workers are trained to identify signs of a possible violent outbreak in the complex social fabric of the neighborhood. Like Massoud Amin's vision of a twenty-first-century power grid, the CeaseFire outreach workers have cultivated a kind of proprioception for their own community. In this way, Clark and her colleagues make up a part of CeaseFire's distributed sensor network.

The mood on the street tonight is relaxed; there is a palpable sense of relief from people in the neighborhood because CeaseFire just successfully mediated a conflict between two warring cliques. Community members can now walk outside their homes without fear of getting caught by a stray bullet. The peace gives the groups of gathering people on the streets a sense of solidarity. Inside the car, the outreach workers chat casually. It's early in the night and they have many hours ahead of them, but the weather tonight is mild—high summer temperatures are another infamous trigger for instigating conflicts. There is every reason to believe that tonight will be "cool."

Suddenly Clark rolls down the window of the moving car. There, on one of the corners, standing with some of the other early evening partiers, is Davion.

"What are you doing out here on the corner?" Clark asks.

Her voice is teasing but kind. The boy comes over to speak to her and he shrugs his shoulders and gives her a charming, capricious smile.

"I'm not doing anything. You know that."

"You better not be doing anything."

"I'm just hangin' with my cousin," the boy says, and he gestures vaguely toward the group of guys lingering on the corner. In the background, the deep reverberations of a bass can be heard.

"Get off this corner! You hear me?" Clark's voice is joking, but there is the slightest register of tension in it.

Davion only looks at her and laughs. He gives her an affectionate wave and turns back to join his group of friends. Clark reluctantly rolls her window back up and starts driving again. "He's got one foot in and one foot out. Just a few more months. Just a few more months and he'll be out of here but . . . Bam! Something can always take them down and keep them trapped."

Clark looks back one more time. In the misty distance, Davion grows smaller and smaller until he is only a blur through the foggy windows.

Before the car turns away down a side street, she takes stock and makes a mental note. Like a good sensor, she will alert the CeaseFire system of a point of vulnerability in the form of a certain boy named Davion. She will not stand for any system failure under her watch.

In the summer of 2008, 125 people were shot dead in Chicago—about double the number of American soldiers killed in Iraq during the same period. These fatalities were just a small part of an epidemic of street violence that is frequently "casual in character," according to Clark. Guns are everywhere. Men shoot one another in disputes over women, because they feel they have been dissed, or—whether they want to or not—because they feel it's expected by their peers. One shooting frequently leads to another, initiating an escalating cycle of violence and retribution that can send neighborhoods careening.

CeaseFire, a campaign under the auspices of the Chicago Project for Violence Prevention, is attempting to reverse this dismal trend permanently. The effort uses an innovative strategy to prevent disputes from tipping over into actual violence, and when and if a shooting occurs, they interrupt any subsequent retaliation, giving the whole neighborhood a chance to cool down. Unlike the intervention designed for the Bangladeshi villages, CeaseFire exists within a state of dynamic disequilibrium, constantly reorganizing itself to suit the needs of its communities. There is no end to this intervention, no final day when all the outside consultants pack up their bags and leave. With CeaseFire, the very people monitoring the community are the exact same people who live there.

In 2000, Gary Slutkin was ready to begin a new project. An epidemiologist by training, Slutkin was returning home to Chicago after stints with the World Health Organization in East African countries like Malawi, Uganda, and Somalia, where he had worked on some of the world's most difficult pandemics. He knew he wanted to move back to the States, but he had no idea what he would work on.

"I really just bumped into this problem of urban violence. It was dominating the neighborhoods. I kept asking people, what is the strategy for dealing with this? There wasn't one."

There are two prevalent framing devices for the problem of violence, Slutkin explained. "People often speak about the need for punishment, including longer prison terms and tougher law enforcement, or people will say, 'You have to improve education, poverty, parenting, and on and on and on.' In public health, we call this the Everything Myth. Punishment doesn't stand a chance because it doesn't drive behavior, and the Everything Myth is just another way of saying that the problem is intractable. It's framed as a choice between doing things that don't work or doing nothing."

Slutkin's previous experience treating epidemics allowed him to see the problem in an entirely new way.

"Malaria was completely resistant to large-scale strategies until bed nets. And reducing diarrheal disease was seemingly intractable with a big Everything Myth attached: 'You need to improve all the water, sanitation, nutrition, etc.' But then they came up with oral rehydration and we have seen enormous improvement. You come up with the right intervention, tie it to the right behavior change, and things really start to move. I saw parallels with violence as a behavior and I figured we could do something using the basic behavior change methods that we already knew. I knew that much but I had no idea what it would look like."

Slutkin spent five years with his staff going through the appropriate steps for designing a behavioral change strategy for violence in Chicago. He sent members of CeaseFire in Chicago to Boston to study what was then referred to as "the Boston Miracle," a successful violence prevention campaign involving innovative partnerships between law enforcement, city officials, and members of the Boston inner city ministries.

"I had experience starting from scratch like this before. When designing the initial strategies for a new public health crisis, we would frequently start with very little information. That would force us to ask basic questions: What is behind this? How is this happening? How is it spreading? This was the exact same feeling. It reminded me of San Francisco when we really had to start at square one."

On June 5, 1981, the Centers for Disease Control (CDC) released the first official document on the disease that would later be known as AIDS. The *Morbidity and Mortality Weekly Report* described five cases of *Pneumocystis carinii* pneumonia, an opportunistic infection typically correlated with a suppressed immune system. All five of the patients were young men—homosexual and previously healthy—living in Los Angeles. By the time the report was out in print, two of them were already dead.

Slutkin was working at San Francisco General in 1981. One month after the CDC's first documentation, a member of the staff saw the hospital's first AIDS case: a twenty-two-year-old man with a rare skin cancer, Kaposi's sarcoma, resulting in lesions across his chest. By the end of that year, nine more people in San Francisco had passed away.

"It was mostly gay men, and we knew that gay men had different sexual practices than heterosexual couples, but why weren't women getting it at all? It wasn't intuitive yet."

Slutkin's group started to do some cluster epidemiology, a time-consuming process that involved interviewing at-risk community members and working with sample and control groups. By 1983, electron microscopes revealed that there was a virus. Less than three years later, more than 800 cases (1 per 1,000 people) had been reported at San Francisco General.

"Suddenly we started seeing some Haitians, some hemophiliacs, some drug users. It started to look like hepatitis B or C, but we still had no idea how it was transmitted. We knew we needed to screen blood, but we also knew that behaviors needed to change."

In the 1980s, during the time that Slutkin was working on strategies for the exploding AIDS epidemic, behavior change in public health was often viewed through the lens of theories like the health belief model. This psychological model was developed in the 1950s as a means of understanding public participation—or lack thereof—in health screenings and prevention programs. The theory focused on attitudes and beliefs of the at-risk population and, as adapted to HIV/AIDS public outreach, it became a framework for understanding the sexual risk behaviors and the transmission of HIV/AIDS.

One of the great limitations of the health belief model, however, was its inability to incorporate the social norms and peer pressure that are so important to consider in any kind of public health behavior. Think of risk homeostasis discussed in the previous chapter. Culture matters. For this reason, strategies in public health were augmented by the theory of reasoned action (TRA), a framework positing that humans are rational and linked in a system of behavioral beliefs (their own attitudes) as well as normative beliefs (the influencing attitudes and behaviors of those around them). Such norms are expressed explicitly and implicitly all around us. Think of the norms dictating the lives of the Bangladeshi women and girls. Of course it would have been easier for them to go to the green well near the mosque, but social norms, informed by religion and gender, made this an unthinkable act.

Even closer to home, we can use the norm of smoking indoors in Western culture. Fifty years ago, societal norms—the cognitive cues signaling appropriate behavior—suggested smoking was an acceptable behavior in most social situations in the United States. And so people smoked everywhere and all the time, as much because they had social expectations of doing so as because they wanted to.

Yet today, smokers in the United States abstain from smoking behavior in almost all common social situations. They are guided by explicit laws and regulations that will fine them for such behavior, for sure, but the real change is one in social norms, which provide a much more powerful deterrent. The risk of a scolding look from a friend or colleague is much more likely than even the most horrific antismoking ad to keep smokers from lighting up. The perceived risk of social death is far stronger than the perceived risk of actual death.

Because they provide powerful cues for which behaviors are acceptable and which are not, an understanding of social norms played a strong role in shaping public health measures surrounding HIV/AIDS prevention in the early days of the disease.

In Thailand, for example, peer use was the greatest factor in determining whether or not a man would use a condom. Much like Cialdini's experiments on hotel towels showed, if your friends used one, you would too. Among a different community, the norms might be very different: In American universities, for example, it was individual attitudes that gave the strongest indication of whether young female co-eds would insist on condom use from their partners. Without a firm understanding of these different communities and their different local norms, public health measures targeting an outbreak would not only be ineffective, they might make the problem worse.

When Slutkin first approached the problem of urban violence, he assumed he would be using a strategy similar to the ones he developed attacking HIV/AIDS, tuberculosis, and cholera over the last twenty years. After sorting through what was working in other cities and what was technically and financially feasible in Chicago, Slutkin and his group created CeaseFire's intervention to tackle the transmission of violence. It can be stated in three simple tenets:

> *First, interrupt the contagion of violence.*
> *Second, change the thinking of the most at-risk transmitters.*
> *Third, change the norms of the community as a whole.*

STEP ONE:
INTERRUPTING THE CONTAGION

Elena Quintana, director of evaluation for CeaseFire, is sitting in her office, a room on the third floor of the University of Illinois/Chicago Department of Public Health building. The hallways outside are lined with mud-green lockers, and all of the offices inside are drab with the speckled floor tiles so often found in institutions. Quintana, by contrast, speaks with impassioned bursts of energy. She has been with CeaseFire since its very inception, coming to work with Slutkin after more than a decade of efforts in domestic violence outreach and earning a PhD in community psychology. Quintana has held a number of different positions within CeaseFire, using her academic training to bridge conversations between the various players. Today she serves as the main conduit between the CeaseFire workers on the street and the Chicago law enforcement. By virtue of her mainstream connections and social network, it is safe for her to be seen walking in and out of the police precinct every Monday. She has no street-level reputation to protect. There is no one in her community watching to see if she is snitching. For that reason, Quintana has become CeaseFire's only public face to the Chicago Police Department.

Yet despite these and countless other efforts by the organization to establish trust, around 2004, the CeaseFire staff realized that they were not reaching deeply enough into the community of at-risk clients. In order to stave off the violent outbreaks effectively, they would need to recruit people who could actually wield insider influence in the neighborhood social networks of its most at-risk citizens.

"You hear people around CeaseFire talk about 'the game,'" Quintana told us. "It basically refers to the way marginalized people are systematically excluded, forcing them to live through an alternative economy: 'the hustle' or 'the game.' It involves all of the illegal activity you hear about: dogfighting, prison, drugs, gangs, bootlegs, prostitution. They were born into it; they're not a part of mainstream society.

We needed to use the people already in the game, use their Rolodex to help clients build a bridge to legitimacy. The system is set up to be us versus them. That is a chasm that most marginalized people just can't cross by themselves, and the outreach workers couldn't span that social distance."

In one of the biggest rooms on the CeaseFire floor, a meeting of highly trained specialists who do span that social distance every day is getting under way. It's a gathering of the CeaseFire violence interrupters, a group comprised mostly of African American and Latino men ranging in age from early twenties to fifties, many formerly incarcerated themselves, who are working on the front lines of the violence epidemic.

Quintana begins the meeting by reading off a police report of neighborhood incidents that occurred over the last week. The men sit and listen and nod their heads, a few commenting under their breath at one detail or another from Quintana's report.

When the report is finished, Tio Hardiman, director of CeaseFire Illinois, stands up to speak to the group. Hardiman, an African American man in his forties, sports a beige tracksuit with navy stripes up the side. He gesticulates smoothly as he speaks to the men; he has their rapt attention.

"Group A has been making trouble with Group B and we're hearing that a shooting is set to happen at eight a.m. We all know that's school time. Kids walking to school then."

The men around the room nod their heads.

Hardiman reminds them of a concept he calls "the rules of engagement." "Our biggest ally is time, remember. We're just borrowing time to calm them down."

Analogous to the Red Team U skeptics, the interrupters cruise the streets of the toughest neighborhoods to identify and intervene in conflicts before they intensify, providing the alternative conceptual space. If a shooting has occurred, they immediately seek out the victim and his or her friends, relatives, and associates and try to prevent a retaliatory shooting, interrupting the spread of violence and stopping it from rippling through the rest of the social network. If CeaseFire is treating violence as a communicable disease, then the violence interrupters function as its public health workers, triaging new infections and inoculating the community to lessen the likelihood of an outbreak.

"I came up with the concept of identifying a specialty unit of guys that come from pretty tough backgrounds, guys that are part of the hierarchy on the streets," Hardiman explains. "These are the guys that have enough backbone to go and actually talk to the guy with the gun in his hand. A lot of guys will talk a good game but not many people can actually go and talk to the guy with the gun in his hand. That's a special skill."

Hardiman plays an essential role, if not *the* essential role, because of his ability to recruit and communicate with these highly influential members of the community.

"Most people would actually like to be talked down from committing a violent act if they are talked to by the right person. It has to be the right person though. If you've got some fluke type of person who shows some weakness, the guy is going to step on you. I've had to mediate conflicts right here in my office and the guys want to get loud and jump up, 'F this and F that! I'm gonna get that chump. That's just the way it goes!' And I say, 'No, that's not the way it goes. You got your girlfriend at home; you got two kids on the way; your mother's sick and now you're talking about shooting someone. What is it about?' And you get down to the nitty-gritty and it's about some guy who owes you one hundred bucks or some guy who messed with another guy's girlfriend or some guy who might just be messing with you, looking at you crazy and trying to intimidate you. So I say, 'C'mon man, let's sit and talk about this. Let's get to the bottom of this.'"

These kinds of interventions are the result of training in an advanced form of social technology. First, it requires CeaseFire's interrupters and outreach coordinators to keep an up-to-date mental map of the social connections in the community—a map of who runs with whom, how various members of the community are linked together, and where, in the social network of the streets, things are heating up and cooling down.

Because the community is always in flux, the map must be refreshed constantly, with highly sensitive information. If CeaseFire were a traditional law enforcement organization employing people from outside the neighborhood, it would never have the trust, credibility, and access required to do its job. And while not all of the Chicago police force is thrilled with CeaseFire's insistence on confidentiality and neutrality when it comes to illegal activity under their watch, many have come to respect CeaseFire's ability to do things the police simply can't do—or to handle things that enable the police to deploy their force more effectively elsewhere.

Just having a good social map isn't enough, however. CeaseFire workers also need to understand the logic of violence at the street level, how an act will be interpreted, not only by the victim but also by dozens of constituencies. In the complex fog of partial information that follows in the wake of a shooting, when every minute matters, where do you intervene? Interrupters need a sense that if X shoots Y, then X's brother will shoot Z, causing Z's friend to escalate—it's scenario planning for the streets.

And then there's the art of the intervention itself—a skill as specialized as talking a suicidal person off a ledge. "It requires real wisdom, authenticity, and a jazzlike set of improvisational skills to talk someone down who has the intent, means, emotional state, and social reinforcement to commit a killing," says Slutkin, speaking of the interrupters. "These guys are masters of a craft that until recently didn't exist."

Like the batfish and the WIR, the violence interrupters function as a countercyclical strategy built into the system, dormant until the social system approaches or passes the critical threshold of a shooting, and then immediately dispatched to recalibrate the network back into a semistable state.

Given the work of the outreach coordinators and violence interrupters, CeaseFire doesn't bear a lot of resemblance to a traditional social service bureau, with office hours and appointments between nine a.m. and five p.m. It's a round-the-clock operation that is most active at night. There are other differences too: Most traditional organizations are opposed to hiring formerly incarcerated workers to do community outreach. CeaseFire, in contrast, depends on them.

"Most organizations work in a more mainstream way," Quintana concurred. "CeaseFire is taking people from the belly of the beast to go back out there and transform lives. It's biblical. You'll hear them say, 'Jesus was an outreach worker.' The violence interrupters often frame their decision to work with CeaseFire in terms of personal redemption."

STEP TWO:
CHANGE THE THINKING

Frank Perez, director for outreach services for CeaseFire, wears a big gold cross around his neck. His black T-shirt reads "Don't Shoot." Perez is of Puerto Rican descent, and he grew up in the violent gangs of South Side Chicago. After years of skirmishes with the law, he had a revelatory moment that changed his life.

"I realized in my late twenties that all this, this whole broken community, this was all my mess. It was our mess and we had to find a way to clean it up. This violence? It's my fault. My whole life now is about trying to undo the damage done."

Perez went on to get a master's in social work before joining up with CeaseFire in 2002. Tonight he is traveling with Kobe Williams through the Englewood neighborhood of Chicago. Williams is a violence interrupter, and Englewood is his beat. Despite the intense nature of his work, Williams seems genuinely at ease. He likes maintaining his connections and getting a chance to spend time on the street. He wears a white baseball cap adorned with the CeaseFire motto, "Stop Killing People." A CeaseFire dog tag swings around his neck.

In the 1990s, Chicago finally made the decision to tear down its infamous Robert Taylor housing project, located in the Bronzeville neighborhood. The Taylor residents were given Section 8 housing cards, which were supposed to allow them to move wherever they wanted to go. In reality, most of the former residents lacked both the social network and financial resources to go anywhere, and the majority simply moved to the next neighborhood over: Englewood. The influx of displaced people disrupted the neighborhood and further exacerbated the problem of fractionalized, violent cliques.

"You used to be able to speak to a gang chief and he would tell his soldiers what to do, but after RICO [Racketeer Influenced and Corrupt Organizations Act] in the seventies and eighties, they got rid of all the gang leaders and put them in jail," Perez explained. "No one wants to step forward and be a leader today. It's too dangerous, too hot. The cops will see you or you will get shot."

Perez feels that RICO—an attempt to empower prosecutors—was a folly of unintended consequences. "They cut off the heads but they didn't kill the bodies, and now there are all these cliques, renegades. They don't respect anyone." Much as John Arquilla describes the task of U.S. forces in wars in Iraq and Afghanistan, CeaseFire is struggling to negotiate with the sense, scale, and swarm abilities of these at-risk potential offenders.

Englewood feels quiet in the early evening. Williams points to one of several empty lots and describes the most recent CeaseFire barbecue held there. "Real popular events. Sixty to seventy people turn out for the free food and they always ask us to do more."

"We call them midnight barbecues," Perez offers. "When an area is hot, we come out at nine p.m. and serve hot dogs as long as we can."

Often the gang and clique members will go about the routine transactions of their drug business on the same corners where CeaseFire is having their barbecues, an example of the trust and authenticity that the CeaseFire strategy wields within the community.

"When we see the gang members selling drugs, we just tell 'em, 'Hey, we're here because you guys are making it too hot with your violence. Cool down.' They know we're right," says Perez.

"In traditional public health, we treat a lot of diseases, like H1N1 influenza, for which we don't yet have effective mediations," Slutkin says. "In their absence, in the early days of dealing with a pandemic, we try to instill certain practices and change behaviors to decrease transmission. And then when we finally do get the vaccine, or antibiotic, we immediately try to get it to people who are most susceptible. This is a strategy based on the understanding of the unique characteristics of the virus and how it's transmitted—in the case of H1N1, it's by coughing and droplets. It's not transmitted by bad people: It's transmitted by air."

And this is where Slutkin pauses in thought for a moment before speaking. He is about to tread into unchartered territory.

"And violence is transmitted by thought."

For Gary Slutkin, violence is not like a disease in the metaphorical sense. Violence *is* a disease, transmitted not by germs but by thoughts, decisions, and ideas—about the nature and acceptability of violence, about the social expectations of one's peers, and about one's sense of hope or hopelessness about the future. It's transmitted from mind to mind—a mimetic, rather than a microbial infection.

And it's extremely contagious.

Many times, the violence that CeaseFire interrupters and outreach coordinators must contend with isn't emotional—it's a calculated necessity, a way to ensure one's status on the streets is not being compromised. What's underlying the violence isn't aggressiveness, or a desire for vengeance, but an idea about how an individual will be perceived by his social group if he doesn't act in the wake of some perceived—or impending—insult.

This fact provides just the opening that interrupters often need to diffuse a conflict before it turns deadly. According to Perez, despite the stereotypes of raging, out-of-control gang bangers, many of CeaseFire's clients "not only come willingly when asked, they often ask for mediation themselves."

Why are these at-risk individuals volunteering to be babysat while they cool down?

"Most of the time," Perez explains, "they don't want to do the deed, but they need to save face. With CeaseFire around, they can tell their friends, 'Oh yeah, Frank talked me down. I would've done it if Frank hadn't talked me down.' And then their friends say, 'Oh yeah, Frank, man, he's good at that.' And everyone gets to walk away with their dignity."

Perez compares it to the fight at the playground: You get in a scuffle during the day with some kid and you set the fight to begin at 3:15 or 3:45. Then the time gets closer and you're getting more and more scared but you have to follow through. "You gotta save your name." So you go out there and all the kids are cheering you on and there's no way to get out.

CeaseFire gives people a way out. Everybody can feel cool.

"I would've done it too, if it hadn't been for CeaseFire."

STEP THREE:
CHANGE THE NORMS

Tio Hardiman's car pulls up alongside a gathering group of mourners on the 3500 block of West Sunnyside Avenue in the Albany Park neighborhood of Chicago. A group of women are crying into crumpled Kleenex. Some of the men are sitting in lawn chairs along the sidewalk, looking and then not looking at the impromptu memorial set up under one of the trees.

ALWAYS REMEMBERED, NEVER FORGOTTEN

I LOVE YOU MY ANGIE

R.I.P. MY NIGGA

The event is honoring Angelina Escobar, nineteen, shot dead in her apartment alongside her boyfriend, Alex Santiago. Hardiman doesn't think that the shooting is directly related to gang or clique violence in the West and South Side neighborhoods, but he is here, representing Cease-Fire and offering solidarity and support in the campaign against violence.

Hardiman approaches the women, Escobar's aunt and other extended family, standing next to the memorial site filled with flowers and handwritten notes.

"My name is Tio Hardiman and I'm here from CeaseFire to show you our support."

Around the CeaseFire offices, you constantly hear the phrase "Credible messages from credible messengers." Hardiman is handing out CeaseFire cards and signs to the growing group of mourners. The message written on them is only two words long: "Don't shoot."

"We're not interested in trying to stop drug use and drug dealing and all of that other stuff," Hardiman tells us. "We have a message that everyone can get behind: Stop shooting people. It's something that everyone agrees on. That's what gives CeaseFire its street cred."

Hardiman points out several members of the CeaseFire Albany Park branch who are now standing in the crowd in solidarity. Before too long, one of CeaseFire's partners in the clergy arrives, the Reverend Robin Hood, pastor of Redeemed Outreach Ministries in the Englewood neighborhood on Chicago's South Side. He stands in front of the group and offers up his hands.

"Another senseless act of violence plaguing our communities. I didn't know Angelina Escobar but I know that she is someone's granddaughter, niece, daughter, friend."

One by one, the mourners move forward to speak about Escobar, an innocent bystander caught in an act of violence. As the last mourner finishes, the reverend repeats his earlier phrase as an incantation: "Another senseless act of violence plaguing our communities." He ends with a CeaseFire chant and the entire crowd joins in: "Stop the shooting! Stop the violence! Stop the murder! CeaseFire! CeaseFire! CeaseFire!"

The words echo through the streets. Like Ury's third side, this public performance of a simple credible message from a credible messenger, over and over again, is intended to broadcast and reinforce a new norm: Shooting is unacceptable. It is the community calling out and responding to the community.

Like transforming the social acceptance of smoking, changing a norm is a long-term process, and it will require all of the tools in CeaseFire's public engagement toolbox, from midnight barbecues and street marches to door-to-door canvasing and dozens of other shared rituals yet to be invented. The effort will succeed only with constant repetition, deep and authentic engagement, sustained effort, and tolerance for inevitable setbacks. And long after the final shooting has plagued the neighborhood, those norms must still be continuously reinforced, to make sure violence never returns.

Behavior change is never simple, but recent research suggests that new norms may spread through a community's social network in much the same way as the mimetic disease of violence itself.

Using data from the famed Framingham Heart Study, an ongoing survey on residents of the town of Framingham, Massachusetts, that has been in continuous operation since 1948, researchers at Harvard University, led by biophysicist Allison Hill, recently found evidence that positive and negative emotions behave like infectious diseases, too, and spread across social networks in a community over long periods of time. Their research found that for each contented person you know, your likelihood of being contented rises 2 percent. Unfortunately, for each discontented person you know, your chances of being discontented rises 4 percent—in other words, unhappy people are twice as damaging to your state of mind as happy people are good for it. Intriguingly, however, they found that happiness "infections" may be more durable—they seem to last twice as long (a decade) as unhappiness infections do.

For CeaseFire, such research suggests not only that normative change is possible, but that the community members they reach indirectly are at least as important as those they reach directly. If violence is a disease spread by thoughts, then perhaps it can be cured by them, too.

Long before that happens, CeaseFire will have made some of the most significant impacts on gun violence ever achieved. In the year 2000, for example, CeaseFire's approach led to a 67 percent drop in the amount

of shootings in the worst police district in the city—the eleventh police
district, the West Garfield Park neighborhood—in a single year. After
receiving additional resources, CeaseFire applied its approach to three
more neighborhoods. By splitting the money across all three communities, it achieved between 33 and 45 percent drops in all the neighborhoods. When it received funds for a fifth neighborhood, the shootings
went down by 45 percent there. CeaseFire moved the number up to
fifteen neighborhoods and then twenty-five neighborhoods and soon
after it was witnessing a feedback system across all of the communities. Today, CeaseFire has achieved a more than 40 percent reduction in
shootings throughout all of its neighborhoods in West and South Side
Chicago.

All of this rests on the outreach coordinators and violence interrupters, who act like macrophages in an immune system, constantly
identifying and containing threats and ensuring that the contagion of
violence doesn't spread through the larger substrate of the community.

Interventions like CeaseFire's are never static. They require buy-in
and support from the community, time to work, and constant vigilance.
But, done right, these kinds of interventions build something extremely
beneficial, which transcends the particular context of violence: a powerful latent social network.

As we will see in the next chapter, leaders with a particular skill set
are then able to tap into this network, amplifying the benefits of the
best-designed interventions across their diverse range of constituencies.

8

THE
TRANSLATIONAL LEADER

As we traveled the world researching this book, one of our most genuinely surprising—and surprisingly consistent—findings was the important role that certain kinds of leaders play in shaping community resilience. This was not our intention: We had not set out to write *The Seven Habits of Highly Resilient People*. And yet, when we discovered a community that was able to reorganize dynamically in the face of disruption, we frequently encountered the same character over and over again, in guises young and old, rich and poor, male and female. These leaders demonstrated an uncanny ability to knit together different constituencies and institutions—brokering relationships and transactions across different levels of political, economic, and social organization.

Such characters don't conform to typical notions of what a leader looks like: They weren't strong-jawed, visionary CEOs or coiffed elected officials, boldly directing from the top, nor were they bottom-up, street-level organizers. Instead, they represented a neglected third form of "middle-out" leadership, seamlessly working up and down and across various organizational hierarchies, connecting with groups who might otherwise be excluded, and translating between constituencies. The authority of these *translational leaders* was not rooted solely in their formal status but in their informal authority and cultural standing. (Think of Josh Nesbit and Patrick Meier in Mission 4636; Tio Hardiman with CeaseFire; or Willie Smits in Indonesia.) When disruption strikes, the presence—or absence—of such a leader can have a profound impact, as we'll see in the island community of Palau, and the remarkable story of one of the translational leaders shaping its future.

UNDERSTANDING PALAU

To visit the island nation of Palau, we took a flight from New York City to Honolulu, traveling five hours back in time. From there, we hopped on a plane bound for Guam, an unincorporated territory of the United States situated far west of the Hawaiian Islands. On the way, we crossed over the International Date Line, going from five hours behind to twenty hours ahead in time. After that, we waited amidst Japanese backpackers and adventure divers to catch the once-daily flight leaving from Guam to the archipelago of Palau in Micronesia.

At the end of our travels, after landing in this exquisitely situated marine ecosystem, we had managed to visit today, yesterday, and tomorrow all in one fell swoop—a blurring of past, present, and future that, as we'll see, could not be more appropriate.

Palau is an archipelago made up of 340 islands in the Pacific Ocean, approximately six hundred miles east of the Philippines. Clustered in the southern half of the islands are the renowned Rock Islands, a series of enchanting rock sprouts popping out of the ocean like lush, overgrown Chia Pets. A barrier reef surrounds the archipelago, the source of the islands' famous lagoon, up to twelve miles wide, shallow in certain areas but reaching 130 feet deep in others. Three major ocean currents converge on the islands, and they bring in all variety of marine life from across the Indian and Pacific oceans. Palau Trench, about 25,000 feet deep, creates upwelling currents that deliver its nutrients to the shallower waters.

The result of all this movement is an astounding diversity of life: The archipelago supports seven hundred coral species—four times as many coral as in the Caribbean.

All of this makes Palau a biodiversity hot spot and earns it a place alongside the Great Barrier Reef as one of Seven Underwater Wonders of the World. It is little wonder that Palauan fishermen are considered some of the most knowledgeable in the world.

Before the twentieth century, these fishermen lived by an age-old conservation ethic that revolved around reef and lagoon tenure. The right to fish was limited by individual communities, and outsiders were forbidden to fish without permission. All of this was legislated through local village chiefs, a role passed down through family lineage. Through this traditional economy, Palauans existed in a state of subsistence affluence—island culture placed a great deal of value on social stability, community support, and family lineage, while the accumulation of physical goods conferred little, if any, status.

Throughout the eighteenth and nineteenth centuries, Palau bounced between the British, Spanish, and German empires—tiny, overlooked, and largely unchanged. But when the islands were officially awarded to Japan following World War I, Palau's traditional culture began to erode. The Japanese incorporated the islands and made them an integral part of their empire. Suddenly Palau's capital city of Koror was transformed from a subsistence economy fishing village into a sophisticated outpost of Tokyo—modern built homes with glass windows and shingled roofs were erected alongside carefully manicured trees and plants, lending Koror's main street an aura of tidy Japanese order. By the 1930s, so many Japanese had immigrated to the islands that native Palauans were the minority in their own country.

With the introduction of the new Japanese culture came a new economic order: The Japanese introduced new nets, motorized boats, and other tools that greatly increased its fishermen's catch. Gone were the days of giving fish away to other villages. Fish were now seen as vital assets to be bought and sold. A market economy was set in motion.

When the United States took over the islands as a trust territory following World War II, these trends accelerated. Research scientist Robert Johannes spent sixteen months with Palauan fishermen in the early 1970s, documenting the elders' recollections of the shifting fishing culture in his pioneering ethnography *Words of the Lagoon*:

> Increasingly he [the Palauan fisherman] found himself forced to compete with his fellow fishermen for money and thus for fish. He abandoned the leaf sweep, which employed a dozen or more men who worked cooperatively and shared their catch, and he adopted the imported "kesokes" net, which he could operate by himself.

In addition to encouraging competitive market-driven behavior and its attendant technologies, the U.S. trust government implemented a centralized democratic governance structure. In 1955, a legislature was chartered and the elected members from the various municipalities were given voting rights. In the transition, the hereditary role of the village chiefs, including their role in fisheries conservation, ebbed. In their place, American-trained civil servants encouraged principles of private ownership, competition, and self-interest. A narrative of national progress provided little incentive to enforce conservation laws: Modernization seemed well worth the cost of a few extra fish.

A tipping point arrived in 1959 when an export market for reef fish opened up in Guam and the U.S. government began to offer the Palauan fishermen loans in order to acquire bigger and more efficient boats. Under pressure to pay off the loans, the fishermen were now even more in need of a plentiful catch. Unsurprisingly, fishermen living in the heavily populated districts around the central islands of Koror and Ngeremlengui quickly started to see their stocks decline. In centuries past, this would have triggered ancient conservation methods to allow the stocks to replenish—setting aside the essential no-take areas that give the fish time to spawn—but in the 1960s and 1970s, the fishermen had no such cushion of time. Instead, representatives from the affected areas began to enact legislation invalidating traditional lagoon and reef laws. They wanted access to all of the fish all around the island.

At the same time, other forces were remaking Palau. Throughout the 1970s, young people started leaving their villages and moving to Koror seeking the lucrative jobs in the civil service, whose salaries were based on the comparatively inflated pay scales of civil servants on the U.S. mainland. This urbanization created its own gravitational force: By 1980, more than 60 percent of Palauans lived in the capital, many of them unemployed. Those with a job could hardly afford the expense of a fishing boat.

In a generation, in the islands making up one of the richest and most diverse systems of marine life on the planet, formerly village-dwelling Palauans found themselves urbanized, underemployed, in debt, and too poor to eat their own fish. The majority of the population subsisted on tinned mackerel salted and shipped in from Japan.

Like many such transitions, Palau's transformation from a traditional, self-sufficient society to an interconnected participant on the global economic grid was a mixed bag. Market-oriented democratic governance brought freedoms and opportunities, to be sure, but it also brought a mix of new incentives and dependencies that eroded long-standing cultural institutions, norms, and taboos. This is not to suggest that precolonial Palau was an unremitting paradise, or that modernity is without abundant virtue. Like all such societies, Palau has permanently crossed a one-way threshold to modernity, for reasons voluntary, seductive, and forcible, and no one but the Palauans are entitled to weigh the costs and benefits. But one fact is inarguable: For all of the extraordinary benefits of global connectivity, the process of modernization also left cultural fissures that fractured traditional institutions, weakened shared values, and threatened the nation's relationship with its extraordinary natural inheritance.

Palau's sudden connectivity with the global economic network illustrates, just like the current fate of the orangutans in Borneo, how frictions can emerge when forces with different time signatures start acting (and interacting) on a complex social and ecological system. Imagine the time it takes to conduct an individual transaction in a Palauan market—the sale of one fish, say, can take about five seconds. But now think about the social and political structure put in place to organize how that fish lands in a fisherman's net. Such a cultural system is put in place over decades—generations, even. Let's scale back even further to ecological and geological time. How many millennia does it take for an ecosystem to evolve enough diversity to sustain that species of fish? And how many more to replace the lost diversity once its population has tipped over into collapse?

Without compensatory mechanisms (such as those being designed by Willie Smits in Samboja Lestari), rapidly moving processes (like market transactions) layered on top of very slow ones (like ecosystem recovery) can erode the adaptive capacity of the system as a whole. Ultimately, the shearing effects can trigger collapse. Precolonial Palau was a place without significant economic dependencies outside of its own archipelago, certainly without a vast amount of material flows coming in and out of the islands. The time signature of society was slow and relatively stable, with a tight coupling between social systems and ecological systems, in a way that provided accurate feedback loops for making conservation choices. But when new extrinsic dependencies were introduced, a new time signature for Palau's economy and value system followed. How does a tiny island nation like Palau navigate such dramatic shifts in priorities and values? When globalization creates more and more dependencies reaching farther and farther around the world—decoupling the time signature of ecological systems from the tempo and rhythm of human societies—how do we reconnect with feedback from the natural world?

What Palau needed was a way to navigate between the ancient and the contemporary, the fast and the slow, the local and the global—to reinforce deeply rooted traditions and build bridges between them and the newer strictures of globalization. Today, with the help of a translational leader, Palau is finding just such a balance.

NOAH AND THE *BUL*

When Palau gained its independence from the United States in 1978, a heady period of change for island government followed. During the transition toward independence, a young native-born Palauan took over as officer of Fisheries Management in the Palau Natural Resources Divisions. Noah Idechong, then in his twenties, exemplified the next generation of Palauan leaders: Although he graduated from Hawaii Pacific University with a degree in business administration, he was born and raised in a small fishing village on the eastern coast of one of the Palauan islands. His formal education in Hawaii allowed him to recognize the importance of administration while also respecting the quickly vanishing knowledge of traditional fishing and the ineffable value of the fisherman's instinct.

Greater ties to the Japanese economy after independence brought the first trickles of tourism. Although the island was long known as a scuba diver's paradise—the drop in depth and changing currents make Palau's dives some of the best in the world—few people outside of the diving community had ever heard of the place. Then, in 1985, a five-star resort opened its doors on one of Palau's private pristine beaches. Suddenly, word spread and, for the first time in its history, Palau needed to negotiate a substantial flow of tourists.

Almost immediately, visiting divers and fishermen found themselves embroiled in conflict. The best fishing spots were also some of the most popular dive spots, and fishermen were fighting for space with divers from around the world. The two groups brought dramatically different frames of reference to the encounter. The divers—often environmentally minded transplants only temporarily engaged in Palau's community—wanted to protect the diversity of the reef; the fishermen were immediately suspicious of outsiders with an agenda for their islands and little apparent appreciation for their often severe financial imperatives.

The conflict between the two groups chilled into a cold war, then escalated when some of the fishermen began dynamiting parts of the reefs in an effort to increase the catch. Both groups brought their grievances to Idechong's desk. What originally began as a position charged with managing fish stocks was now growing into something ever more complicated: a position charged with managing people.

"We only knew how to build a fishery," Idechong said. "We never thought about things like stakeholder management and sustainability. The only framework we ever used was maximum sustainable yield. That was how fisheries of the past were run. I just assumed we were a part of that past."

Desperate to quell the increasing conflicts, Idechong looked to his staff at Marine Resources for a solution. "We had nobody. We had no scientists, no officers, no one. All we had were people who knew how to catch fish and how to catch fishermen. That changed my attitude. We needed to figure out how to manage a fishery with a more diverse group of people."

Idechong decided to intervene in the conflict in two stages: To begin with, he would create a dialogue with the fishermen. He hoped to convince them to work together with the divers, shifting more of the island's economic focus from fishing toward an ecoconscious form of tourism. In the second phase, he would engage with divers and other members of the tourist communities and work with them to create a green fee, or tourist tax, that would benefit the Palauans, rewarding them for meeting their conservation goals.

To achieve this ambitious, two-phase strategy, Idechong needed to start by reinvigorating the ancient practices of conservation. He assumed that he could simply go to the chiefs and ask them to reassert their power over their individual reef systems. When he initiated a conversation with these village leaders, however, he consistently heard the

same complaint: "The national government has taken away that power."

Idechong was surprised by this response. While the power of the chiefs had demonstrably diminished, for sure, there had been no formal retraction of the traditional lagoon and reef tenure policies. The erosion of their power came not from any official legislation but, rather, from a kind of learned helplessness on the part of the traditional leaders. The proliferating market forces, which had gradually pulled resources and influence into the national government in Koror, had spread the perception that all power and resources lay in the capital. The traditional leaders, though nominally recognized, had become little more than figureheads. With no institutional muscle behind them, village chiefs were regarded as a quaint, if denuded, relic.

And yet, the new government had also failed in its attempts at governing land tenure and ecosystem management. Idechong recognized that if Palau had any hope of sustaining its vital resources, a combination of the two authorities—traditional and contemporary—would be needed.

"I went back to Parliament and said, 'We have no plan, no money, no solutions, no scientists. Nothing,'" Idechong said. "If we are going to do this, we are going to have to invent local, Palauan solutions."

One of the most extreme practices of conservation in Palau is traditionally referred to as the *bul*, a moratorium on fishing for an explicitly stated amount of time, protecting spawning channels for the fish, which allows them to replenish themselves. Idechong felt that only this drastic measure of conservation could begin to bring back the diminishing fish stocks: a win-win for both the fishermen and the divers. But he also knew that the chiefs, now rendered all but powerless, would be unlikely to risk further humiliation. So he came to them with an unusual idea. Idechong offered to serve as legal counsel for the local villages, helping them navigate the often-alien language of official government.

After they agreed, he spent the following months researching the Palau constitution side by side with the traditional laws. He discovered that though the central government had taken away the local right of

the chiefs to fine, there were still many vagaries in Palauan law surrounding legislation of the *bul*.

"I saw some light there. I said, 'You guys do the *bul* and, if something goes wrong, I'll try to support you from the legal end.' It wasn't perfect, but we had no other choice."

The first *bul* went into effect in the early 1990s at various locations throughout the islands. Almost immediately, village chiefs from the communities in northern Palau ran into difficulties asserting their enforcement authority. Idechong realized he needed to create legislation at all scales—from local to national—to mirror the traditional practices, offering them the scaffolding of legal language.

"Mirroring was a critical part of the puzzle," Idechong said. "We asserted traditional practices and authority, and we asserted formal legal authority at the same time. When we employ a *bul* at the national level, we mirror that with a local initiative at the community level—enforcing the protection from all angles."

It worked. Idechong's first mirror legislation, enacted in northern Palau, was so successful that the local government and the village chiefs worked in concert to make the whole area a preserve for perpetuity, with limited fishing takes and an educational component.

From that moment forward, whenever a *bul* was imposed, Idechong set up a local government law instituting no-take areas for three years, mirroring the three-year *bul* moratorium on fishing in the area. After more than a decade of dialogue with local fishermen and village chiefs, in 1994, Idechong was able to mirror an entire network of traditionally deemed no-take areas all around the island with a national law titled the Marine Protection Act.

The next part of the solution involved getting the fishermen to work with the divers to create adaptive capacity for an economically viable tourist industry. Idechong used his credibility with the local people to articulate the benefits of bringing more outsiders to Palau.

"I said, 'Here is a fish and his name is Herman. If you take him out

and sell him at the market, you might get seventy dollars. But if the tourists come and look at him over and over again, you're going to make much more money than that.' I showed them how we calculate the amount using the number of dives that the tourists make in any particular site and then extrapolate the value of the fish for the given year. The value of the tourist fish is always much, much higher. When the fishermen understood that, they got on board. They wanted to become boat operators and tour guides because they could see the benefits of tourism in a tangible way."

The final stage in Idechong's strategy asked something of the divers and the other tourists starting to visit Palau. He convinced Parliament to institute a green tax on the tourists to help keep money coming in to protect the MPAs (Marine Protected Areas).

"In the old days, you would conserve and immediately bring the benefit into your village. These days, the local people ask, 'Where are our benefits?' The rewards go directly to the national government because they are able to reach their biodiversity and conservation goals while the villages do all the work. They complain that the water is serving the tourist industry, the police force is safeguarding the tourists, the infrastructure is put in place to make the tourists happy. I say, let's move some of the tourist money directly to the community. There is no shame in that. Give them the tangible benefits for their efforts."

Today, after considerable effort from Idechong, the tax is officially in place and positively received by both the locals and the divers. In 2012, the minister of finance expects to collect more than $1.5 million from the green fees—an enormous amount for a country the size of Palau—all of which will go directly to support community management efforts on the ground.

"Because the money is flowing, I don't have to deal with grumblings anymore. People realize what it's for and both the tourists and the local people are seeing the benefits. We created a transparent system where the benefits are clear to everyone."

A TRIBE EXPANDED

In 1994, Idechong moved away from his position with Fisheries Management to head up the not-for-profit conservation initiative called the Palau Conservation Society. He felt that working with an entity that existed outside the government structure would allow him to better reach the villagers and engage them in enlarging and connecting the areas in the Marine Protection Act to create a PAN, or Protected Areas Network.

It was during this period that Idechong developed one of his most strategic partnerships. He served as the bridge between the Western marine biology and ecology scientists flocking to Palau and the village chiefs and fishermen he was working with in the Palau Conservation Society.

"I knew I needed to have the right information, the right science," Idechong told us. "But I also wanted to make sure I was working with scientists who knew how to stay out of the way."

Bob Richmond, a researcher in marine conservation biology at the University of Hawaii, has been working with Idechong for almost thirty years. In one of their earliest meetings, Idechong told Richmond with full candor what he wanted from him.

"He said, 'I don't need you to come in and tell the people of Palau what to do,'" Richmond explained to us. "Noah thinks that a good partner is a scientist who comes in with the most reliable information. Then he feels it is his job to translate that information to the stakeholders."

Implicit in this translation process is Idechong's drive for outcomes as opposed to outputs. As Richmond put it, "Lots of scientists and researchers in my field and in the government are counting the number of workshops, the number of people attending, the number of posters created. Idechong, by contrast, is looking at what is happening at the reef level: Are the fisheries improving? Are the corals coming back? Is the ecosystem resilient to both human and natural stressors?"

In 2005, Richmond approached Idechong and the Palau Conservation Society with funding from a research grant. He wanted to study the relationship between people and nature in a Palauan watershed with results that could be of use to the communities served by Idechong's organization.

Richmond and his team of scientists were moving forward on potential sites for the study when Idechong interceded.

"He told us that he wanted a place that would bring about better outcomes for conservation efforts," Richmond explained. "Instead of letting pure science dictate our choice, he chose the watershed that served as the main drinking water source for Koror. It was in an area of mangroves that was rapidly being cleared for housing lots, causing devastation to the reefs. Idechong felt that data from that specific watershed would have the greatest impact on people's behavior, galvanizing them to protect the area."

Richmond and his team went door to door in the watershed area and, with Idechong's guidance, they got complete buy-in for the study from every member of the community.

Next Idechong insisted on hiring local fishermen to check on the gear and equipment that were a part of the study. Initially Richmond assumed this was out of fear that the equipment would be stolen, but Idechong quickly corrected him.

"He wanted eyes and ears in the water all the time. He wanted his guys to go and have a beer with the fishermen and get a day-to-day perspective on details: 'Oh, it rained today' or 'I was out in such and such an area and I didn't see any fish either.'"

His plan worked. By the time Richmond and his researchers were ready to present their collected data to the villagers, the fishermen were already engaged in the project and eager to hear the latest news.

"They knew everything! They were up on every bit of data that we had collected and we never said a word to them about it. There was no learning curve at the presentation because they were completely primed with all the data and on board with the work we were doing. That was just a part of Noah's brilliance: He kept the information open and transparent."

After Richmond and his fellow scientists finished presenting their data to the villagers, Idechong made it clear that they would now sit in the back of the room and effectively become invisible. As Richmond put it, "We were there in a supportive role."

Two young Palauans, Yimnang Golbuu and Steven Victor, both mentored by Idechong and sent on for doctoral training at Richmond's lab in Hawaii, presented their findings on the watershed to the village chiefs.

"We're sitting in this traditional village house and they set up a PowerPoint. All the foreigners stayed in the back, quiet and respectful. Noah made a point of sitting with the fishermen, his people, while the chiefs sat in the front. And then we all listened while two Palauan PhDs presented to fellow Palauans with a Palauan language PowerPoint."

What happened next still astounds Richmond, and he references it as a moment of profound revelation for his work as a scientist and researcher.

"After the presentation, the chiefs were very deferential to Yim and Steven. You could see it in their body language and in their tone: 'Could you explain this' and 'I didn't quite get that . . .' But then something changed. The mood became more confrontational. The village chiefs began using a harsher tone, almost as if they were reprimanding the young Palauan scientists."

Richmond looked to Idechong for reassurance. *Is everything all right?* he telegraphed with his face and eyes.

Idechong returned the look with a signature twinkle in his eye. He nodded his head and just sat back and watched.

"When we came out later, I said to Noah, 'What the heck happened at the end there? That was not a peer-to-peer conversation. It sounded like they were being lectured.'"

Idechong explained that in the hierarchical culture of Palau, transfers of information must be exchanged between the elders and the younger scientists. What the chiefs were basically saying, Idechong told Richmond, was, "You guys came to us and really impressed us as Western scientists. You brought in good information and we are proud and impressed by your work. But any good Palauan would know that the term for sediment on land is different than the term for sediment on the water."

The message, Idechong noted, was, "You guys have come in and taught us some really interesting things from your scientific education and now it is our duty to teach you to be better Palauans."

The dynamic Richmond and his research team observed typifies the traditional social reciprocity that characterizes Pacific island cultures. Getting information without returning some in exchange is considered rude. To Richmond, the village chiefs sounded like they were chiding the young scientists, but they were actually expressing gratitude by giving something back. These are the kinds of subtleties of culture that are often lost without the presence of a translational leader.

"It was actually that information exchange that sealed the conservation deal," Idechong told Richmond.

"What's funny," Richmond mused, "is my colleague at the meeting, a brilliant modeler, was giving us flack about even attending. He kept saying, 'Why are we going to this meeting? Is there going to be a publication coming out of this? Why bother?' And we just kept saying, 'Shut up and watch.' After the meeting, he came out raving. He said, 'I've never seen that kind of dynamic at a scientific presentation to stakeholders before.'"

Idechong's delicate social maneuvering paid off in the policy implementation of the study. Within six weeks of the presentation, the traditional leaders approached the elected leaders and demanded a moratorium against development in the watershed. It went into effect within six months of the presentation and remained valid for several years. That bought Idechong enough time to legislate through the state and federal level to insure protection of the coastal mangrove in perpetuity. When the moratorium ended, the mangroves were officially protected under national law as a marine protected area, conserved in the language of tradition as well as the language of law.

Today Noah Idechong is an elder statesman of Palau. We met him for breakfast at one of his many haunts in Koror, the Penthouse Hotel. Although he has the same mane of dark hair and curious sparkle in his eyes, the Idechong of today is too busy to hold court for long. In the early 2000s, he returned to government, winning election as a speaker of the Palau National Congress (constituting both the House of Delegates and the Senate). While serving in the legislative branch, Idechong has been a tireless advocate for his conservation agenda at both the national and international scale. He convinced then-president Tommy Remengesau to try scuba diving for the first time. The experience converted Remengesau, and he made conservation a keystone issue for his term, signing into law a total ban on shark finning, deep-sea bottom trawling, and the live reef fish trade in Palau as well as approving Idechong's Protected Area Network Act, legislation that ensures long-term financial sustainability of the marine network.

These days, Idechong divides his time between Palau and the other island nations in Micronesia, working to replicate and scale his original success in new contexts. He is talking up his Micronesia Challenge initiative (a commitment by all Micronesian nations to effectively conserve at least 30 percent of the near-shore marine resources and 20 percent of the terrestrial resources across Micronesia by 2020), brokering funding for it with local and international conservation agencies like the Nature Conservancy, Conservation International, and Global Environment Facility. He is also bringing the message of the *bul* and integrative leadership to the United Nations.

"Other people offer solutions," Idechong told us, "while I offer dialogues. When I work with people, we discuss and we discuss and I let them think and then we discuss some more. I provide assistance to guide them in their thinking, but there is no end to the process. There is no set goal."

Idechong has honed this process after years and years of seeing outside organizations come into Palau and promise quick fixes through attempts at infrastructure improvements and community building projects.

"The project gets money and then, when the project completes, the entire effort dies and everyone disappears. This is the kind of thing I have steered away from. I adopt my partnerships for a lifetime."

Conservation scientist Michael Guilbeaux has worked with Idechong for many years and he considers him both a colleague and a mentor: in his words, "a Palauan father."

"Noah is extremely open about incorporating the ideas from science but he wants to do it in an appropriate way for Palau. There is a healthy circumspection about the science. In the face of climate change and some of these other global issues, what can we do? Noah's feeling is that science can only take us so far when we're dealing with social systems. Scientists can study this thing to death but Palauans need to get down to business and do their work."

And what is that work? we asked him.

"The work is building capacity. All people can do is build adaptive capacity."

Creating that capacity required Noah Idechong to seek out the correspondences between the ancient and contemporary; between the scientific and the social; between the ecological and the economic; between formal government and informal governance. He achieved this by using the central tool of all translational leaders: brokering relationships between organizations, constituencies, and forces operating different levels of scale, organization, and time within a system.

Translational leaders do not dispense with hierarchies; they recognize and respect their power. Instead, standing at the intersection of many constituencies, translational leaders knit together social networks that complement hierarchical power structures. Rooted in a spirit of respect and inclusion, these complementary connections ensure that when disruption strikes, all parts of the social system are invested, linked, and can talk to one another.

WEAVING THE NETWORK

The activity at which Idechong so naturally excels is also referred to as "network weaving," a term coined by social network analysts Valdis Krebs and June Holley, based on their extensive work exploring how to build resilience in rural communities in Appalachian Ohio. And although he might not have done so explicitly, Idechong was demonstrating several of network weaving's core principals.

Krebs and Holley describe how a resilient community network emerges through four stages: First, small, autonomous clusters emerge, often without any guidance, among individuals and organizations with shared interests, values, and goals. In the Palauan example, this might be represented by the close connections between the commercial fishermen or the reef divers. These clusters serve to reinforce interest politics, and if their interconnectivity ends there, these groups can remain oppositional and the larger social structure weak and brittle to disruption.

In the second and more intentional stage of network weaving, translational leaders like Idechong create a hub and spoke model, with themselves as the initial hub, connecting many different kinds of constituencies. Doing so often requires a mixture of charisma and grace, and the knack for navigating the politics of difference. During this second phase, translational leaders spend much of their time learning about the network they're building, discovering what each of its nodes knows and what each needs. Authenticity and an ethic of generosity are critical at this stage, as the network has a single point of failure—the leaders themselves. But if done in the proper spirit, the emerging network—and the leader at its hub—will grow a reputation as a connector and begin to develop its own gravitational force. Idechong's early work bringing together divers and fishermen correlates broadly with this second phase.

In the third phase, translational leaders begin to close the triangles in their network—building direct bridges between different constituencies for whom they are the sole bridge. This starts to create a multihub, or "small world," social network.

Due to the number of relationships involved at this point, the best network weavers don't just connect—they teach those they connect how to become connectors themselves. "This transition from connector to facilitator is critical," Krebs told us. "If the change is not made, the network weaver at the center can quickly become overwhelmed with connections, and the growth and efficiency of the network slows dramatically—or can even reverse course." At this point the translational leader must quickly change from being a direct to an indirect leader, guiding the emergence of new network weavers throughout the community.

If successful, a multihub network forms and a new dynamic also emerges—the power of weak or indirect ties, particularly between hubs in a social network. These provide vital bridges between groups with different perspectives or expertise, or they may evolve into strong ties themselves, bringing hubs closer together. Idechong's effort to build direct, unmediated bridges between, for example, Bob Richmond's scientific research team and Palauan fishermen is an example of this stage of network weaving.

The final stage of Krebs and Holley's model, and its ultimate aim, is called a *core/periphery social network*. In this highly stable yet highly resilient social arrangement, which usually emerges after years of effort, a core of strongly affiliated hubs at the center of the social system is connected to a constellation of people and resources on the periphery, through weak ties. This allows for an efficient and natural division of labor: The periphery monitors the environment, while the core implements what is discovered and deemed useful. "The periphery allows us to access new ideas and new information from outside—the core allows us to act on them, inside," says Krebs. This kind of core/periphery model is exactly what Idechong is building today, as he connects strongly functioning cores across Micronesia to learn from one another.

This is not to say network weaving is a magic bullet. The capacity that translational leaders create does not belong to them—it belongs to the community itself, and there's never a guarantee of its sufficiency in the wake of disruption. Nor does network weaving eliminate competitive forces at work in a community. There will always be oppositional people and organizations at work, and translational leaders cannot pretend such oppositions don't exist. Instead, they can look for points of collaboration within the larger competitive reality.

"In a healthy network, you have to connect on your similarities and compete on your differences," says Krebs. To illustrate, he tells the parable of the jars: "Imagine a community in which two women make and sell jam—one sells organic jam, the other sells exotic flavors. Both are competitors, and neoclassical economic theory suggests they should compete ruthlessly. But they are also community members and neighbors, and they need to have a relationship for the social network to remain healthy. Perhaps there are ways for them to cooperate in a limited way: for example, to aggregate their purchase of jars so that they each lower their costs. This won't eliminate their competition, but it will create the opportunity for a relationship with a broader definition and a hint of mutualism. Often, before you weave the network, you have to find the jars."

As you can see, at various times, translational leaders must be connectors, mediators, teachers, behavioral economists, and social engineers. They must carry out these duties with candor, transparency, generosity, and commitment. They must also embrace key principals of social network creation: Build your network before you need it. Build direct relationships so that, in a pinch, reconfiguration and collaboration can emerge quickly, but not so many relationships that things become densely overconnected. And most important, create the context, trust in the participants, and know when to let go.

9

BRINGING RESILIENCE HOME

In our tour of resilience patterns, we have crossed a range of geographies, disciplines, contexts, and ideas. We've seen some of the ways systems can become fragile and break. We've examined how their connectedness can create feedback loops that (often invisibly) either amplify or erode their resilience. We've explored how resilience strategies from one domain might find application in another. And we've glimpsed how individuals, groups, and communities can bolster their resilience by embracing a warm zone of connectedness, collaboration, and diversity.

It turns out that Goldilocks had it right all along: Resilience is often found in having just the *right* amounts of these properties—being connected, but not too connected; being diverse, but not too diverse; being able to couple with other systems when it helps, but also being able to decouple from them when it hurts. The picture that emerges is one of strategic looseness, an intentional stance of both fluidity (of strategies, structures, and actions) and fixedness (of values and purpose).

The problem, of course, is that many of the aspects of a given system's resilience are defiantly context specific. The particular approach that makes one organization more resilient in a given situation may make another more fragile. (Note the word "more"; there are no absolutes in resilience, no binaries, just measures of more and less.) Every resilient solution is rooted in its setting and not necessarily a sure path to others' success.

How, then, do we put these principles into action?

MAPPING FRAGILITIES, THRESHOLDS, AND FEEDBACK LOOPS

While there's no single recipe for every circumstance, every journey toward greater resilience begins with continuous, inclusive, and honest efforts to seek out fragilities, thresholds, and feedback loops of a system— grasping its holistic nature, identifying its potential sources of vulnerability, determining the directionality of its feedback loops, mapping its critical thresholds, and understanding, as best we can, the consequences of breaching them. Doing so calls us to greater mindfulness—assessment, without judgment, of the world as it truly is. And this is just as true for organizations and communities as it is for people.

As we've seen, fragilities can take many forms: chronic, long-term challenges like persistent poverty; or the erosion of social mobility; or increased susceptibility to environmental shocks; or the vulnerability of supply chains and infrastructure to climate disruptions. Fragilities can emerge when a corporate culture slowly migrates to an inappropriate level of risk tolerance; or when effective governance wanes; or when a loss of cognitive diversity leads to groupthink; or when a loss of biodiversity devastates an ecosystem. Fragilities can be rooted in the impossibly complex Gordian knots that sometimes tie systems together, as we saw in the tortilla riots, or in the brittleness of a system that is unable to cooperate when it counts. In an organization or a community, fragility can arise from a lack of trust among community members or employees, from a deep resistance to change, or from the slow erosion of psychological resilience. And fragilities don't even have to be negative, per se: They can take the form of a "golden handcuffs" dependency on an external subsidy, or on a single mode of highly profitable but undiversified operations.

All civilizations face their fragilities. Many residents of the world's wealthiest nations, particularly Americans, have felt fortunate to live through a period largely insulated from shocks and disruptions. This "vacation from history" enabled many to become accustomed to living at the efficient-but-fragile end of the robust-yet-fragile continuum. In a world temporarily devoid of consequences, the slow erosion and increasing inelasticity of the country's political, financial, socioeconomic, and ecological systems scarcely seemed to matter. Now that a new, more volatile chapter has begun, those now-compromised systems have flipped from being engines of resilience to sources of fragility themselves.

All of these vulnerabilities are distinct and require different responses, yet all have the same effect: They make their host systems less adaptive and increasingly robust-yet-fragile—capable of dealing with anticipated disruption but likely to collapse in the face of an unorthodox challenge.

In spite of this, surprisingly few communities or organizations have any kind of structure in place to think broadly and proactively about the fragilities and potential disruptions that confront them. This has to change. Today, it's unthinkable for an organization of any meaningful size not to continuously monitor its financial or supply chain risk; soon, it will be equally unthinkable not to scan for a broader array of potential disruptions, from environmental issues such as carbon, water, energy, and climate risks to internal cultural factors like levels of cooperation and trust, and social issues such as the health and well-being of the communities these organizations operate in.

Even so, it's important to remember that seeking fragilities does not guarantee finding or eliminating them. Surprises are by definition inevitable and unforeseeable, but seeking out their potential sources is the first step toward adopting the open, ready stance on which resilient responses depend.

This is not to say that resilience is solely about risk management—in the decades to come it's also going to become a significant driver of innovation in its own right. Developing new technologies and services that help enterprises decouple themselves from scarce ecological resources or that help customers contend with unpredictable shocks is going to be big business, worth many billions, if not trillions of dollars.

Consider Kilimo Salama, an innovative agricultural microinsurance program developed in rural Kenya, and one example of a resilience strategy with significant future economic prospects. Created by the Syngenta Foundation for Sustainable Agriculture, UAP Insurance, and Kenyan mobile operator Safaricom, Kilimo Salama (which means "safe farming" in Swahili) insures tens of thousands of smallholder farmers, who may plant as little as a single acre of corn, against the possibility of severe weather events such as drought and excessive rain. These are precisely the kinds of events that are likely to become more frequent in a future increasingly influenced by climate change.

Kilimo Salama's customers are often near-subsistence farmers, whose labor costs are very low but whose adaptive capacity in the face of bad weather is almost nonexistent: One badly timed drought can be catastrophic. To mitigate this risk, Kilimo Salama enables farmers to purchase very small insurance contracts on each bag of seed they buy, at the point of sale, for 5 percent of the cost. Contracts are purchased using M-Pesa, a mobile payments platform deployed ubiquitously throughout Kenya, on the farmer's own mobile phone.

Once they've registered a farmer, Kilimo Salama uses a series of distributed, solar-powered wireless weather stations to monitor climate patterns around the farmer's land throughout the growing season. If there's too much rain, or too little, at the wrong time, the farmers get a payout automatically via their mobile phones, covering the cost of their seeds.

The innovations here are numerous: First, using automated weather sensors means that one of the costliest aspects of delivering insurance—having to check manually if an adverse event has actually affected a particular farm—is effectively eliminated. "Farm visits make sense if you have a large-scale farm, because the cost of the visit can be rolled into the policy," says Rose Goslinga, Kilimo Salama's founder. "But for a small-scale one-acre farm, the math just doesn't add up—and this has been among the main reasons insurance hasn't been available. By using wireless sensing stations, we don't have to visit the farm, which transforms the economics."

Similarly, by linking the weather sensors directly to the M-Pesa accounts of the customers, the entire claims process is eliminated: If the weather turns bad, payment is automatically made via a mechanism the farmers know and trust, much faster than if a claims adjuster had to be involved. Better still, the information that these sensors collect is used to develop insights on relevant regional climate trends. In a positive feedback loop, these are then delivered back to farmers via targeted, timely text messages helping them to become better farmers even during good years.

Not only does microinsurance insure against loss, it lets people make investments they wouldn't feel comfortable making otherwise, because they can hedge them. Goslinga points to one farmer she knows who took a chance on a higher-yielding, more expensive variety of corn only because he could ensure it with Kilimo Salama. His yield increased 150 percent.

Other innovations are less technical but no less important to the success of the program. Developing Kilimo Salama required Goslinga and her colleagues to figure out how to use the technology to bundle and service nontraditional customers in a nontraditional way, how to restructure the insurance supply chain accordingly (all the way up to the reinsurers, who insure the insurance company), and—most important—how to get skeptical farmers to sign on.

The resulting service is not just an important part of the future of climate resilience, it's going to be an important part of the future of the global insurance industry itself. The MicroInsurance Centre estimates the market for microinsurance products of all kinds to be at least 1 billion people, less than 3 percent of whom are currently served. With opportunity like that at hand, the Kilimo Salama model will soon be exported, not just to other countries in the Global South, but, in a reversal of received stereotypes about the flow of innovation, also to countries in the Global North. In this century, resilience will not just be a social good, but a global market opportunity from which developed nations have as much to learn as they do to teach.

EMBRACING ADHOCRACY

As we've seen again and again, wherever we find social resilience, we rarely find just big, formal institutions at work. Instead, we often find a rich stew made up of bits and pieces of public and private organizations, informal social networks, government agencies, individuals, social innovators, and technology platforms, all working together in highly provisional, spontaneous, and self-organized ways. Since each disruption and circumstance is unique, there can be no prefabricated organizational chart for the players—indeed, one of their first tasks is to create it.

There's a name for this mode of organization, which was first popularized in the 1970s by the futurist Alvin Toffler and management theorist Henry Mintzberg: It's called an *adhocracy*. It's characterized by informal team roles, limited focus on standard operating procedures, deep improvisation, rapid cycles, selective decentralization, the empowerment of specialist teams, and a general intolerance of bureaucracy. In the digital age, an adhocracy can be put together in a plug-and-play, Lego-like way, well suited in fast-moving, fluid circumstances where you don't know what you'll need next. If it were a musical genre, adhocracy would be jazz.

A successful adhocracy was, for example, at the heart of the success of Mission 4636 initiative in Haiti, in which dozens of self-coordinated collaborators—some as large as the International Red Cross and the U.S. Marines, others as small as an individual graduate student volunteer—worked together with nothing more than shared purpose and good software holding them together. Adhocracy also underwrites the success of CeaseFire's violence interrupters, who are able to bring together the right array of people and resources, at the right time, to cool down a shooting. In these cases and others like them, the resilient response is improvisational and provisional, established at the speed circumstances dictate.

In a larger sense, adhocracy is what all of the lessons, insights, and strategies in this book are really about. Whether expanding the cognitive diversity of a team, like Greg Fontenot and the leaders of Red Team University; or weaving diverse social networks together like Noah Idechong in Palau; or facilitating cooperation among traditional antagonists like William Ury and his colleagues at Abraham Path; or co-opting the swarming tactics of al-Qaeda in the Arabian Peninsula; or promoting eBay-like approaches to managing complex campaigns as espoused by John Arquilla; or creating complementary currencies like the WIR; or building the adaptive capacity of individuals through mindfulness training like Elissa Epel and her colleagues—all help to ensure that an adhocracy can emerge and thrive when it's needed.

The converse is also true: When systems are structurally overconnected—as they were in the 2008 banking crisis; or when responses fail to represent a diversity of genuine stakeholders, as they did at 33 Liberty Street; or when interventions are bureaucratically imposed on communities rather than developed with them, as in the case of the Bangladeshi wells, there is no space for an adhocracy to germinate.

This is not to say that formal institutions don't have a role in resilience—they certainly do. But when we focus too strongly on them as the *sole* actors in response to a disruption, we don't just ignore, but can actually smother the opportunities for these kinds of successful, improvisational approaches to emerge. Unfortunately, in fields ranging from international development and diplomacy to disaster recovery, that's exactly where our instincts take us: We usually bias in the direction of the bureaucracy, rather than away from it. What's needed is an approach that complements these silos of excellence and works in the white spaces between them, where resilience (and social innovation) is so often found. That's what resilient organizations, and their translational leaders, do: They create the opportunity, connectivity, permission, and encouragement for people to meet in the white spaces. The leadership imperative in such circumstances is centered on influence and coordination, not command and control.

THE FIERCE URGENCY OF DATA

Adhocracies thrive on data. And by a stroke of fantastic luck, we're currently witnessing the global birth of an adhocracy of data—a global revolution that, for the first time, empowers organizations with the capacity to collect and correlate widely distributed real-time information about the way many critical systems are performing. This kind of open data will play a central role in resilience strategies for years to come.

A perfect example is given by FLOW, a powerful new data reporting system developed by the international water and sanitation organization Water for People, which is used for measuring the success of water and sanitation projects over time.

"Today, we rightly celebrate the day when a community gets water for the first time," says Ned Breslin, Water for People's CEO and a longtime veteran of global water projects. "But the way many water and sanitation agencies operate, it's as if this is the end of the story. It's not—it's just Day One. Unfortunately, due to poor planning and a lack of community ownership, within a few years of being completed, up to 60 percent of the water projects in Africa and Asia are broken or abandoned. I've walked right past broken wells dug five or ten years ago to build new ones."

To tackle this problem, Water for People developed FLOW, or Field Level Operations Watch, software that allows field-workers to use their mobile phones to map the health of water access points—sort of an Ushahidi for water. The system provides at-a-glance reports on thousands of health and sanitation points in a given region—not only their operational status, but also the level of community support and oversight—and injects a new level of transparency, efficiency, accountability, and resilience into water and sanitation efforts. Imagine how differently things might have turned out in Bangladesh had such a platform existed. Or in New Orleans after Katrina, for that matter.

Mining the data exhaust from platforms like FLOW and Ushahidi can tell us a lot about where systemic fragilities are emerging and where interventions aren't working—often in a matter of minutes, rather than months. By making this kind of data externally available (with appropriate safeguards for privacy and anonymity) it can be remixed and correlated to produce dramatically more valuable insights, far beyond what might be feasible for any one organization to collect. That in turn helps separate an early, weak signal of a genuine disruption from simple background noise, and gives us more time to act upon it.

In the future, this data will increasingly be the fodder for sophisticated, predictive machine-learning algorithms that will not only help identify where weak signals are occurring, but where they will likely occur, helping us not only to react faster, but preemptively.

That's the theory behind EpidemicIQ, an initiative to track the global emergence of public health epidemics worldwide through the collection and correlation of literally billions of social media data points every day. The system monitors countless Twitter feeds, Facebook updates, mobile text messages, blog posts, local news reports, and the like, in thousands of languages around the world, in real time. A sophisticated artificial intelligence system ingests and sorts through this gargantuan mountain of data, seeking to separate possibly relevant reports from irrelevant ones. Once potential hits are identified, they're sent to human beings for verification, including, ingeniously, online players of the video game *FarmVille* on Facebook. (Players are offered virtual goods in exchange for correctly identifying whether a particular social media post is actually about a disease outbreak.) These confirmed hits are then sent on to a pool of subject-matter experts who analyze them using proven epidemiological models. This is an adhocracy of a different sort: of artificial intelligence software, untrained human beings, and skilled experts, all seamlessly collaborating to find digital needles in vast digital haystacks.

And it works. During a recent outbreak of Ebola in Uganda, EpidemicIQ was tracking cases five days faster than the Centers for Disease Control and eight days ahead of the World Health Organization. But the ultimate promise of such systems may be in telling us where the pattern might go next, before it does. Cross-referenced with data about human and animal migration, transportation systems, trade networks, and countless other variables, a future fast-moving contagion might be arrested by selectively decoupling and islanding parts of various global systems without inducing mass panic. Imagine how a similar system of financial observatories might help regulators like Andy Haldane in the next economic crisis.

Such possibilities are a big part of the reason global organizations like the UN and the World Bank have made commitments to opening the data about their development programs around the world. It then becomes possible to correlate outcomes from, say, the UN's food relief efforts with the quality of water access points from Water for People and the development programs funded by the bank. This is also why global companies like Nike, GE, and IBM are creating open data initiatives covering many aspects of their operations. They're betting that the competitive cost incurred by becoming more transparent will be dwarfed by the value of understanding how their own operations relate to other critical global trends.

These approaches also hold promise for gluing governments and informal community networks together for greater resilience. An effort called Open311, for example, seeks to create a standardized interface to the 311 public information service used in many American cities like New York, so that a resident could, for example, automatically report a pothole, receive a text alert when its filled, and map a route to work that is as pothole free as possible. Such a gateway could be used for far more than reporting when something is broken; it could also collect subjective data on how residents feel about the safety and livability of their neighborhood, or residents' thoughts on what to do with a vacant lot, or how to guide volunteer efforts during a crisis to the places they are most needed. These efforts expand resilience by broadening the possibilities of cooperation, collaboration, imagination, and responsiveness among the city's residents. In so doing, they expand the varieties of informal governance far beyond what the formal government could do on its own—once again, by embracing adhocracy. And for organizations of all types there is a powerful lesson here: Resilience benefits accrue to organizations that prioritize the collection, collation, presentation, and sharing of data.

This is not to say that the open data revolution is without risks or downsides: Just as correlations can be made to sense and respond to trouble, they can be used by those seeking to create trouble as well. Some governments—and not just despots and dictators—are already finding open data tools an irresistible tool for tracking citizens in ways that may violate their constitutional liberties. And it's not just governments, but criminals and terrorists who are also getting into the game: After Mexican drug cartels successfully silenced their traditional media by killing journalists who reported on their activities, they took to reverse engineering the identities of bloggers and social media users as well, recently hanging several of their mutilated corpses from a bridge with a sign reading, "This is what happens to people who post funny things on the Internet. Pay attention."

Equally disturbing, during the Mumbai hotel terrorist attacks mentioned in chapter 2, the terrorists on the ground remained in constant satellite phone contact with a distant coordinator, who looked up the online identities of hostages; when he confirmed that a hostage was, for example, an American or Jewish, his compatriots shot him. In a further perverse echo of Mission 4636, the same coordinator scanned Twitter postings from the terrified populace and used it to improve the attackers situational awareness and body count; indeed, the open data was so valuable to the terrorists, the Mumbai government asked the populace to stop tweeting live updates about the attack.

There are no simple answers here. Crisis mappers and those involved in the humanitarian community have begun to discuss ways to improve security and to implement codes of conduct and shared principles for harm minimization, but these will always be in tension with the need to ensure that platforms for sharing open data are as accessible and participatory as possible. Ultimately, like many aspects of our modern information society, it comes down to a value judgment: that while the risks of potential abuse are real, they are, in the main, vastly outweighed by the benefits.

REHEARSING THE FUTURE

Even with robust data in hand, healthy communities can't be expected to anticipate *all* of the surprising, nonlinear ways in which systems behave when they're disturbed, any more than the experts do. What's needed is an inclusive way of thinking about various possible, probable, and preferable futures and their implications. Here again, new tools and technologies promise to help communities and organizations map critical thresholds, rehearse the consequences of disruption before they occur, and make better choices.

One pioneering example is a scenario-planning platform called Marine InVEST, developed by a consortium called the Natural Capital Project. Like an ecologically focused mash-up of Google Maps and the video game *SimCity*, the Marine InVEST software allows coastal communities to develop a sophisticated perspective of the many interactions between the land and the sea in their particular geographic region.

Starting with a spatial map of the coast and aquascape, ecologists add science-driven models of local biodiversity and species distribution, ecosystem services, fishing, industrial activity, shipping, recreation, tourism, and the like. The resulting map illuminates the real financial value of the services that the ocean currently provides and makes clear how various activities constrain and shape one another. It also translates between denominations of value in the system—revealing the underlying exchange rate of dollars, fishing stocks, biodiversity, tourists served, miles of coastline preserved, etc.

Many of the connections between these systems are multidimensional. For example, adding fish-farming pens to a system might increase a local company's revenue, but it may also affect wild fisheries, as pens are a perfect breeding ground for parasites that affect penned and wild species alike. At the same time, adding such pens can alter the water quality by increasing effluents, which in turn can affect how much of a coastal system can be used for recreation.

By making visible these kinds of connections, Marine InVEST enables onshore communities to generate plausible answers to real-world what-if questions: What impact might a specific increase in commercial fishing have on nonfished species? How would installing an ocean-energy system affect aquatic recreation and tourism? What's the future cost (and benefit) of protecting a certain amount of shoreline—not just in dollars, but in biodiversity? How much additional activity is safe before important variables are pushed close to a critical threshold?

Understanding the answers helps the community holistically balance competing economic, ecological, and social interests while avoiding the critical thresholds that could flip the system into a less desirable state—just as ecosystems-based fisheries management (EBFM) approaches call for.

Picking a specific approach is no mean feat as, in the real world, some individual stakeholders will inevitably be shortchanged to some degree: Fewer fish will be caught than could be caught; fewer tourists will be accommodated than could be. No plan is without its compromises, and getting people to live with less-than-peak efficiency isn't a software problem—it's a political one.

To address this, the Marine InVEST team works closely with the community to build consensus around the goals, trade-offs, and costs of a given scenario. "Without this engagement, the results are meaningless," says Anne Guerry, one of the project's scientists. "On the other hand, when the community members have scenarios in hand it changes the nature of their conversation. It lets stakeholders play things out in a particular direction, see the risks and rewards of a particular scenario and how it might get into dangerous territory and fail, but without the real-world consequences of doing so."

In a manner reminiscent of Red Team University, this form of serious play encourages community members to think through the complex and less-than-obvious second- and third-order implications of a given set of choices. And it often reveals as much about the nature of relationships between the stakeholders as it does about the implications of the scenario itself. "In many ways, these renewed relationships and engagement processes *are* the software's deliverable," says Guerry. "The models get better every time we engage with nonscientists and"—echoing Noah Idechong's work in Palau—"so do their relationships with one another."

Even a tool as sophisticated as Marine InVEST won't capture every possible source or consequence of disruption. There will always be surprises, and nothing can force people to make the "right" decisions. But getting communities to reflect together about the implications of such scenarios builds resilience nonetheless: A community that learns to discuss one possible disruption is better prepared to deal with any possible disruption.

LEARN FROM RESILIENT PLACES

It's important to state again here that the path to resilience doesn't mean the elimination of every disturbance. As we noted earlier, artificial, prolonged stability can itself be a sign of fragility—an indication that a system may be conserving too many of its resources, like a forest overdue for a prescribed burn. Weathering the occasional disruption is one of the most important ways systems learn, by turning the spotlight on potentially severe fragilities without causing the system to flip completely into a degraded state. They also serve as mechanisms of creative destruction (to borrow Schumpeter's famous phrase) that can clear a path for new arrangements—in our political system, our communities, our ecosystems, our infrastructure, and our economies alike. Modest, regular disruptions also amplify the internal diversity of the system—ensuring that some part of it is continuously seeding, growing, maturing, dying, and fertilizing the whole.

Over time, this messy cycle shapes the culture of resilient places and communities and the beliefs and values of the people who live in them. This may be one reason that social resilience is often found in places that have experienced deep challenges, like the Gulf Coast, the South Side of Chicago, or Detroit. Their all-too-routine, painful experiences of disruption build a deep cultural memory of resilience.

That's the lesson of the last story in this book: the story of Hancock Bank and its remarkable response to Hurricane Katrina.

Thanks to a culture of extensive disaster planning, the 112-year-old bank, based in Gulfport, Mississippi, was able to get its essential computer operations back up and running just three days after the storm ransacked its gleaming seventeen-story headquarters.

But on the ground, in the wake of the storm, that scarcely seemed to matter. Ninety of the bank's 103 branches had been either obliterated or severely damaged; electricity was wiped out over the entire region; police and firefighters were occupied with disaster recovery and unable to provide banking protection; and many customers had lost everything, including their IDs and checkbooks. Amid the chaos, it was impossible for Hancock's staff even to know who their customers were, not to mention how much money they had on deposit with the bank. And yet, with credit-card systems offline, citizens needed cash more

urgently than ever before, and all of the local sources—Hancock's and
its competitors' alike—were devastated.

So what to do?

At the height of the calamity, the bank's executives went back to the
institution's founding charter, more than a century old, and noticed that
it focused entirely on serving people and taking care of communities.

The word "profit" never appeared.

So Hancock Bank's executives did something remarkable—they
enlarged the tribe. While the winds were still subsiding, Hancock employees stood in front of forty of the branches knocked offline, operating from card tables, under tarps, and out of mobile homes, and offered
two hundred dollars in cash to anyone who would sign a slip of paper
with his or her name, residence, and Social Security number. Not just
Hancock customers—anyone. No ID, no problem. Much of this money
was recovered from the ruins of local casinos and literally laundered
and ironed by branch workers before being handed out.

By the time the disaster was over, the company had put more than
$42 million into the hands of local residents—customers and noncus-
tomers alike—with nothing more than Post-it Note IOUs as a record.
This helped keep the local economy going at a time when it faced
months of near-certain paralysis.

Hancock's incredible act of trust paid off handsomely for the bank:
Within months, thirteen thousand new accounts were opened at Hancock and deposits had risen by $1.5 billion; within three years, all but
$200,000 worth of those initial $200 loans—99.5 percent—had been
paid back.

At the center of this remarkable story was a clear, shared vision of
the bank as a long-term social institution first, and a short-term profit-
making institution second. The bank had obviously prepared for disasters, but no one had contemplated handing out millions to strangers,
two hundred dollars at a time. It was Hancock's extensive rehearsal for
future disruption; its strong, shared social values; its trust in its community; and its empowered adhocracy of employees and stakeholders

that fed its ability to rapidly flip, like the batfish or the WIR, into a completely new mode of behavior, in which the normal rules of operation were suspended and both the bank and the community could be restored.

And no regulator had to tell them to do it.

In so many of the stories we've explored about resilience, we see a great confluence of factors coming together—the right systems and structures, the right technologies and information, the right kinds of community empowerment, and the right values and habits of mind.

When these are joined, multiple victories ensue, no matter the scale or context. Neighbors who work together to design and plant an urban farm in a vacant lot, for example, will certainly go some of the way toward reversing urban blight and improving their food security, health, and nutrition. They'll save a little money, get some exercise, and suck a little carbon dioxide out of the atmosphere all at the same time. While they're at it, they may amplify their own sense of self-reliance and adaptive capacity and build the relationships that may ensure their community's resilience in the face of some crisis yet to come.

Even our own thoughts play a role here, not only in our own resilience, but in others' as well. The work of researchers like Richard Davidson, Elissa Epel, Clifford Saron, Amishi Jha, and others shows us paths to improving our own resilience through reflective practice and the discovery of greater meaning in our lives. And Gary Slutkin shows us how such habits of mind can be contagious, for good or ill. Tie these threads together, and you have the first links in a chain that connects your personal resilience to that of your social circle, your community, the place you work and live, and out across the world. What you choose to believe, the mental practices you cultivate, and how you respond to disruption truly shape the whole. Resilience can radiate out from within.

The journey toward resilience is the great moral quest of our age. It is the lens with which we must necessarily adjust our relationships to one another, to our communities and institutions, and to our planet. Even so, we must remember that there are no finish lines here and no silver bullets. Resilience is always, perhaps maddeningly, provisional, and its insistence toward holism, longer-term thinking, and less-than-peak efficiency represent real political challenges. Many efforts to achieve it will fail, and even a wildly successful effort to boost it will fade, as new forces of change are brought to bear on a system. Resilience must continuously be refreshed and recommitted to. Every effort at resilience buys us not certainty, but another day, another chance.

Every day is Day One.

ACKNOWLEDGMENTS

No matter how many authors are on the cover, every book is the work of many hands. We would like to thank the people who made this project possible, starting with our saintly spouses, Jennifer Carlson and Bryan Mealer, who not only supported us in the writing, but also looked after two budding new families at the same time. Our agent Zoë Pagnamenta not only championed this project but also deftly and graciously steered it through several fragile moments on the path to completion. A team of research assistants, including Mishka Vance, Michael Brady, Andrea Jones Rooy, Mirela Iverac, and Jonathan Thong, helped us sort through the science. Dominick Anfuso at Free Press and Hilary Redmon helped us bring it over the finish line. Andrew would also like to thank his wonderful colleagues at PopTech, particularly Leetha Filderman, who kept the wheels on the bus turning when Andrew appeared after a night of binge typing. Andrew would also like to thank his parents. His father's strength and mother's unparalleled ability to bounce back have been guiding stars his whole life.

A special thanks to the remarkable people whose work and writing directly (and as often indirectly) inspired the ideas presented in this book—particularly Buzz Holling, Brian Walker, Johan Rockström, John Doyle, George Sugihara, Simon Levin, Robert May, Andy Haldane, David Bellwood, Bernard Lietaer, John Arquilla, Sarah Fortune, Massoud Amin, Dan Nocera, Geoffrey West, Willie Smits, George Bonanno, Richard Davidson, Clifford Saron, Amishi Jha, Robert Axelrod, Frans de Waal, William Ury, Hijazi Eed, Frederic Masson, Josh Weiss and everyone involved with the Abraham Path in Palestine,

Patrick Meier, Josh Nesbit, Rob Munro, Sinan Aral, Greg Fontenot, Kevin Dunbar, Gary Slutkin and the entire organization of CeaseFire, Noah Idechong, Valdis Krebs, Rose Goslinga, Ned Breslin, Rod Salm and the Reef Resilience scientists working around the world, Charles Allen and the late (great) Pam Dashiell, Ann Guerry, David Rand, and Robert Kirkpatrick. A special thank you to Janet Ginsburg, for her thoughtful commentary on the manuscript. The work of many notable organizations, including the Stockholm Resilience Centre, the Resilience Alliance, the Community and Regional Resilience Institute, Frances Westley and the University of Waterloo Institute for Social Innovation and Resilience among many others, all left an imprint on our thinking. The field of resilience is filled with some of the most remarkable and inspiring thinkers we've ever encountered, and it's an honor to learn from them and introduce them to you.

Finally, a note of thanks from Ann Marie to the iced coffee served "toddy style" at Café Regular in Park Slope. Andrew wishes to thank the good people at Nero Doro café in Brooklyn, New York, for indulging him for hours on end and keeping him caffeinated as he hunched over a laptop.

Abraham Path
www.abrahampath.org

Ushahidi
http://ushahidi.org

PopTech
http://poptech.org

ORGANIZATIONS WE ADMIRE

The world's understanding of resilience is still unfolding, and major leaps are being made every day. Here are just a few of the organizations that are leading the way:

The Resilience Alliance
www.resalliance.org

The Stockholm Resilience Centre
www.stockholmresilience.org

The Community and Regional Resilience Institute (CARRI)
www.resilientus.org

Transition Network
www.transitionnetwork.org

Doors of Perception
www.doorsofperception.com

Ecotrust
www.ecotrust.org

NOTES

INTRODUCTION:
THE RESILIENCE IMPERATIVE

1 *35 cents a pound*: James McKinley Jr., "Cost of Corn Soars, Forcing Mexico to Set Price Limits," *New York Times*, January 19, 2007. www .nytimes.com/2007/01/19/world/americas/19tortillas.html [accessed July 23, 2009].

1 *three months earlier*: "Mexicans stage tortilla protest," BBC News, February 1, 2007. http://news.bbc.co.uk/2/hi/6319093.stm [accessed July 23, 2009].

2 *elite of the country*: Ioan Grillo, "75,000 Protest Tortilla Prices in Mexico," *Washington Post*, February 1, 2007. www.washingtonpost.com/wp-dyn /content/article/2007/02/01/AR2007020100210_pf.html [accessed July 23, 2009].

2 *the 2,900 oil rigs*: Robert L. Bamberger and Lawrence Kumins, *Oil and Gas: Supply Issues After Katrina and Rita*, Congressional Research Service, October 3, 2005. http://assets.opencrs.com/rpts/RS22233_20051003.pdf [accessed November 25, 2011].

2 *for several months*: Elliot Blair Smith, "Katrina cripples 95% of gulf's oil production," *USA Today*, August 30, 2005. www.usatoday.com/money /industries/energy/2005–08–30-katrina2-refinery-usat_x.htm [accessed November 25, 2011].

2 *in a single day*: Kent Bernhard Jr., "Pump prices jump across U.S. after Katrina," MSNBC.com, September 1, 2005. www.msnbc.msn.com /id/9146363/ns/business-local_business/t/pump-prices-jump-across-us -after-katrina/#.TtCVsqNWqUc [accessed November 25, 2011].

3 *cost to produce it*: Timothy A. Wise, *Agricultural Dumping Under NAFTA: Estimating the Costs of U.S. Agricultural Policies to Mexican Producers* (Washington, DC. Woodrow Wilson International Center for Scholars, December 2009), 4.

3 *their Mexican subsidiaries*: Ana de Ita, "Fourteen years of NAFTA and the tortilla crisis," bilaterals.org, www.bilaterals.org/IMG/pdf/fightingFTA -en-Hi-2-h-fourteen-years-nafta-tortilla-crisis.pdf [accessed November 15, 2011].

3 *out smaller suppliers*: Laura Carlsen, "Behind Latin America's Food Crisis," WorldPress.org, May 20, 2008. www.worldpress.org/Americas/3152 .cfm [accessed December 1, 2009].

3 *net food importer*: Walden Bello, "Manufacturing a Food Crisis," *The Nation*, June 8, 2008. www.thenation.com/article/manufacturing-food-crisis [accessed January 15, 2011].

3 *third largest importer:* "China Emerges as the Second Largest U.S. Agricultural Export Market," USDA Foreign Agriculture Service, December 20, 2010. www.fas.usda.gov/China%20Import122010.pdf [accessed January 15, 2011].

4 *unambiguously to climate change*: Miguel Llanos, "2011 already costliest year for natural disasters," MSNBC.com, July 12, 2011. http://today .msnbc.msn.com/id/43727793/ns/world_news-world_environment# .TtxX5eNWqUc [accessed December 1, 2011].

7 *in resilience research*: For a concise explanation, see www.youtube.com/ watch?v=tXLMeL5nVQk [accessed January 15, 2010].

7 *opportunities, resources, and dangers*: C. S. Holling, "Resilience and stability of ecological systems," *Annual Review of Ecological Systems* 4 (1973): 1–24. See also B. Beisner, D. Haydon, and K. Cuddington, "Alternative stable states in ecology," *Frontiers in Ecology and the Environment* 1, no. 7 (2003): 376–82.

7 *existential valley below*: These are commonly referred to as "basins of attraction" in the resilience research community. See also Marten Scheffer et al., "Catastrophic shifts in ecosystems," *Nature* 413 (2001): 591–96. doi:10.1038/35098000.

8 *availability of clean water*: Johan Rockström et al., "A safe operating space for humanity," *Nature* 461 (2009): 472–75. doi:10.1038/461472a.

9   *social media service*: U.S. Geological Survey, Twitter Earthquake Detector (TED). http://recovery.doi.gov/press/us-geological-survey-twitter-earthquake-detector-ted [accessed December 15, 2009].

10   *they are needed*: Nathan Eagle, "Engineering a Common Good: Fair Use of Aggregated, Anonymized Behavioral Data." In press.

10   *impending economic disruption*: Nathan Eagle et al., "Community Computing: Comparisons between Rural and Urban Societies using Mobile Phone Data," *IEEE Social Computing* (2009): 144–50, and personal correspondence with Nathan Eagle.

10   *cotton T-shirt*: Alexandra Alter, "Yet Another 'Footprint' to Worry About: Water," *Wall Street Journal*, February 17, 2009. See also Lorrie Vogel, speech at the 2009 PopTech conference. http://poptech.org/popcasts/lorrie_vogel_pioneering_designs [accessed January 1, 2010].

11   *Hora and Tempus*: Herbert Simon, "The architecture of complexity," *Proceedings of the American Philosophical Society* 106 (1962) 467–482. The authors are indebted to Andy Haldane for this parable, which also appears in Haldane's "Rethinking the Financial Network" speech delivered at the Financial Student Association meeting, Amsterdam, April 2009.

18   *cycle begins anew*: If this idea sounds like Schumpeter's notions of creative destruction, that's because it was based on them. For a good overview, see Brian Walker and David Salt, "Resilience Thinking: What a Resilient World Might Look Like," *Sockeye Magazine*, Autumn 2007.

20   *arrival of climate change*: For an introduction, see www.transitionnetwork.org, the website of the Transition movement, and Rob Hopkins, *The Transition Companion: Making Your Community More Resilient in Uncertain Times* (London: Chelsea Green Publishing, 2011).

21   *necessitating fewer deliveries*: Theatre of the Absurd: Starring Del Monte's Single Plastic Packaged Bananas, http://blog.friendseat.com/del-monte-bananas-single-plastic-package [accessed August 18, 2011].

CHAPTER 1: ROBUST, YET FRAGILE

25   *plot of earth*: For more on this thought experiment, see J. M. Carlson and John Doyle, "Highly Optimized Tolerance: Robustness and Design in Complex Systems," *Physical Review Letters* 84 (2000): 2529–32.

27   *in this case beetles*: For a good overview, see John Doyle, "The Architecture of Robust, Evolvable Networks: Organization, Layering, Protocols, Optimization, and Control," research overview for the Lee Center for

Advanced Networking. http://leecenter.caltech.edu/booklet.html [accessed January 15, 2010].

29 *extra, useless information*: For further elaboration, see John Doyle et al., "The "robust yet fragile" nature of the Internet," *Proceedings of the National Academy of Sciences* 102, no. 41 (2005): 14497–502.

30 *across the network*: Ashley Frantz, "Assange's 'poison pill' file impossible to stop, expert says," CNN.com, December 8, 2010. http://articles.cnn.com/2010–12–08/us/wikileaks.poison.pill_1_julian-assange-wikileaks-key-encryption [accessed January 15, 2011].

30 *coordinated cyberprotests*: John Leyden, "Anonymous attacks PayPal in 'Operation Avenge Assange,'" *The Register*, December 6, 2010. www.theregister.co.uk/2010/12/06/anonymous_launches_pro_wikileaks_campaign [accessed December 15, 2010].

33 *base of the reef*: Richard W. Zabel et al., "Ecologically Sustainable Yield," *American Scientist*, March–April 2003, 150–57.

33 *pounded the reefs*: L. S. Kaufman, "Effects of Hurricane Allen on Reef Fish Assemblages near Discovery Bay, Jamaica," *Coral Reefs* 2 (1983): 43–47. Also J. D. Woodley et al., "Hurricane Allens Impact on Jamaican Coral Reefs," *Science* 214 (1981): 749–55.

34 *long-spined sea urchins*: Office of National Marine Sanctuaries, National Oceanic and Atmospheric Administration. http://sanctuaries.noaa.gov/about/ecosystems/coralimpacts.html [accessed January 15, 2009].

34 *"a single living individual"*: Nancy Knowlton, "Sea urchin recovery from mass mortality: New hope for Caribbean coral reefs?" *Proceedings of the National Academies of Science* 98, no. 9 (2001): 4822–24.

34 *for a marine organism*: Ibid.

35 *drop of 98 percent*: "The California Sardine Industry," Trade Environment Database. www1.american.edu/TED/sardine.HTM [accessed January 15, 2010]. See also John Radovich, "The Collapse of the California Sardine Industry: What Have We Learned?" in *Resource Management and Environmental Uncertainty* (New York: Wiley, 1981).

35 *theories were wrong*: Paul Raeburn, "Using Chaos Theory to Revitalize Fisheries," *Scientific American*, February 2009. See also C. H. Hsieh, "Fishing elevates variability in the abundance of exploited species," *Nature* 443 (2006): 859–62. doi:10.1038/nature05232.

36 *species are overfished*: Boris Worm et al., "Rebuilding Global Fisheries," *Science* 325 (2009): 578–85.

36 *maximum catch levels*: Boris Worm et al., "Impacts of biodiversity loss on ocean ecosystem services," *Science* 314 (2006): 787–90.

36 *fish in the sea*: Ibid.

39 *a true win-win*: James Sanchirico, Martin D. Smith, and Douglas W. Lipton, "An Approach to Ecosystem-Based Fishery Management," Resources for the Future Discussion Paper, DP-06–40 (2006). www .rff.org/Documents/RFF-DP-06–40.pdf and www.rff.org/Publications/ Resources/Pages/Managing-fish-portfolios.aspx [accessed January 15, 2010].

39 *"Ecology for Bankers"*: Robert M. May, Simon A. Levin, and George Sugihara, "Ecology for Bankers," *Nature* 451 (2008): 893–95.

41 *transferring funds between them*: Kimmo Soramäki et al., "The Topology of Interbank Payment Flows," Federal Reserve Bank of New York, Staff Report no. 243, March 2006.

41 *approximately $2.4 trillion*: Fedwire Funds Service annual data. www .federalreserve.gov/paymentsystems/fedfunds_ann.htm [accessed September 1, 2011].

42 *sixty-six banks accounted for 75 percent*: Soramäki et al., "Topology of Interbank Payment Flows."

42 *"too big to fail"*: Ibid.

42 *have increased sixfold*: Andy Haldane, "Rethinking the Financial Network," speech delivered at the Financial Student Association meeting, Amsterdam, April 2009. www.bankofengland.co.uk/publications /speeches/2009/speech386.pdf [accessed May 18, 2010].

47 *1 billion pages*: Ibid.

48 *various market participants*: Ibid.

51 *more than 75 percent*: International Swaps and Derivatives Association. www.isda.org/uploadfiles/_docs/ISDA_Brochure_2011.pdf [accessed July 1, 2011].

57 *under exceptional circumstances*: David Bellwood, Terry Hughes, and Andrew Hoey, "Sleeping Functional Group Drives Coral-Reef Recovery," *Current Biology* 16 (2006): 2434–39.

58 *32 percent in 1932*: Peter Temin and Gianni Toniolo, *The World Economy Between the Wars* (Oxford: Oxford University Press, 2008), 96.

58 *reached 30 million*: www.digitalhistory.uh.edu/database/article_display .cfm?HHID=462 [accessed January 15, 2010].

58 *90 percent of their value*: "Times Topics: The Great Depression." http://topics.nytimes.com/top/reference/timestopics/subjects/g/great _depression_1930s/index.html [accessed January 15, 2010].

58 *seventy-three official job seekers*: Tobias Studer, "WIR and the Swiss National Economy," translated by Philip H. Beard, WIR Bank, Basel. www

.atcoop.com/WIR_and_the_Swiss_National_Economy.pdf. [accessed November 15, 2011].

58 *and increased unemployment*: James P. Stodder, "Reciprocal Exchange Networks: Implications for Macroeconomic Stability," conference proceedings, 2000 IEEE EMS. http://ssrn.com/abstract=224418 [accessed July 8, 2008].

CHAPTER 2: SENSE, SCALE, SWARM

62 *industry trade publication*: *Inspire*, issue 3, November 2010. www.slideshare.net/yaken0/inspire-issue-3 [accessed January 15, 2011].

63 *reading Islamic texts*: Scott Shane and Souad Mekhennet, "Imam's Path from Condemning Terror to Preaching Jihad," *New York Times*, May 9, 2010, A1.

63 *"already faltering economy"* : *Inspire*, issue 3, November 2010.

63 *to mechanical failure*: "UPS cargo plane crashes in Dubai, killing two," BBC News, September 3, 2010. www.bbc.co.uk/news/world-middle-east-11183476 [accessed January 31, 2011].

64 *FedEx facility in Dubai*: "Bomb was designed to explode on cargo plane—UK PM," BBC News, October 30, 2010. www.bbc.co.uk/news/world-us-canada-11657486 [accessed January 31, 2011].

64 *France's interior minister*: "French Minister Says Yemen Bomb Detected 17 Minutes Before Exploding," Voice of America, November 4, 2010. www.voanews.com/english/news/europe/French-Minister-Yemen-Bomb-Detected-17-Minutes-before-Exploding-106689223.html [accessed January 15, 2011].

64 *"we call leverage"*: *Inspire*, issue 3

65 *in their pockets*: Caroline Gammell, "Christmas bomb plot: nine men remanded over plan to 'blow up Big Ben and Westminster Abbey,'" *Telegraph*, December 27, 2010. www.telegraph.co.uk/news/uknews/terrorism-in-the-uk/8227193/Christmas-bomb-plot-nine-men-remanded-over-plan-to-blow-up-Big-Ben-and-Westminster-Abbey.html [accessed January 15, 2011].

65 *mode of conflict*: John Arquilla and David Ronfeldt, eds., *Networks and Netwars: The Future of Terror, Crime, and Militancy* (California: RAND Monograph Reports, 2001).

68 *sick with the illness*: "Tuberculosis," fact sheet 104, World Health Organization, November 2010. www.who.int/mediacentre/factsheets/fs104/en/index.html [accessed January 15, 2011].

68   *die from the disease*: "2010/2011 Tuberculosis Global Facts," World Health Organization. www.who.int/tb/publications/2010/factsheet_tb_2010 _rev21feb11.pdf [accessed January 15, 2011].

68   *reading this sentence*: "Tuberculosis," fact sheet 104.

71   *up to sixty*: Electronic communication with Sarah Fortune.

71   *finally pass away*: "Tuberculosis and MDR-TB," Partners in Health. www.pih.org/pages/tuberculosis-and-mdr-tb [accessed February 20, 2011].

74   *"of the 20th century"*: "Greatest Engineering Achievements of the 20th Century," National Academy of Engineering. www.greatachievements .org. [accessed May 10, 2008].

75   *North American grid operating*: *The Emerging Smart Grid*, Global Environment Fund and Center for Smart Energy, October 2005, 1. www .smartgridnews.com/artman/uploads/1/sgnr_2007_0801.pdf [accessed January 15, 2010].

75   *$160 billion*: Ibid.

75   *high-voltage superhighways*: Ibid.

76   *after 4:00 p.m.*: *Final Report on the August 14, 2003 Blackout in the United States and Canada: Causes and Recommendations*, North American Electric Reliability Corporation U.S.–Canada Power System Outage Task Force Report. www.nerc.com/filez/blackout.html [accessed January 31, 2011].

77   *in eight U.S. states*: http://en.wikipedia.org/wiki/Northeast_blackout_ of_2003 [accessed January 31, 2011].

77   *North American history*: www.semp.us/publications/biot_reader.php?Biot ID=391 [accessed January 31, 2011].

77   *the age of eighteen*: "Blackout Stalls Economy, Transportation, Public Services," Fox News, August 15, 2003. www.foxnews.com/story /0,2933,94795,00.html. [accessed January 31, 2011].

77   *first time in decades*: www.illinoislighting.org/loss.html [accessed January 31, 2011].

78   *"especially during emergencies"*: Massoud Amin and Phillip F. Schewe, "Preventing Blackouts," *Scientific American* 296 (2007): 60–67.

78   *"Maybe it did"*: Ibid.

79   *municipal water supply*: Kim Zetter, "H(ackers)2O: Attack on City Water Station Destroys Pump," Wired.com, November 18, 2011. www.wired .com/threatlevel/2011/11/hackers-destroy-water-pump [accessed November 25, 2011].

79 *their useful lives*: Bart Tichelman, "Using a Smart Grid to Address Our Aging Infrastructure," *Utility Automation & Engineering T&D*, October 1, 2007, 56–56.

81 *to one another*: Martin LaMonica, "Cisco: Smart grid will eclipse size of Internet," CNET News, May 18, 2009. http://news.cnet.com/8301 –11128_3–10241102–54.html [accessed February 8, 2010].

84 *fuel resupply convoys*: *Sustain the Mission Project: Casualty Factors for Fuel and Water Resupply Convoys*, Army Environmental Policy Institute, September 2009.

84 *four hundred dollars a gallon*: Bryant Jordan, "Gas Costs $400 a Gallon in Afghanistan," Military.com, October 20, 2009. www.military.com/news /article/gas-costs-400-a-gallon-in-afghanistan.html [accessed February 16, 2010].

84 *200,000 gallons a day*: Marine Corps Expeditionary Energy website. www.marines.mil/community/Pages/ExpeditionaryEnergy.aspx [accessed February 17, 2010].

84 *make this possible*: Rick Maze, "'NetZero' aims to cut greenhouse gases on bases," *Marine Corps Times*, July 12, 2011. www.marinecorpstimes.com /news/2011/07/military-energy-defense-department-bases-071211w/ [accessed August 6, 2011].

84 *Power Shade*: Kris Osborn, "Army evaluating transportable solar-powered tents," December 8, 2010. www.army.mil/article/49138/army-evaluating -transportable-solar-powered-tents [accessed January 11, 2011].

85 *platoon's command center*: "Solar Energy Powers Marines on Battlefield," press release, Office of Naval Research, December 7, 2009. www.onr .navy.mil/en/Media-Center/Press-Releases/2009/Greens-Solar-Energy -Marines.aspx [accessed February 17, 2010].

85 *at the same time*: Wayne Arden and John Fox, *Producing and Using Biodiesel in Afghanistan*, June 2010. www.biodieselinafghanistan.org/uploads /AFGH-PAPR-20100609-EXEC.pdf [accessed January 11, 2011].

85 *hydrogen and oxygen*: Matthew W. Kanan and Daniel G. Nocera, "In Situ Formation of an Oxygen-Evolving Catalyst in Neutral Water Containing Phosphate and $CO2+$," *Science* 321 (2008): 1072. doi:10.1126 /science.1162018.

86 *who don't have it*: "Tata funded MIT founded startup Sun Catalytix to provide solar power storage for low income houses in India," Panchabuta .com, November 30, 2010. http://panchabuta.com/2010/11/30/tata -funded-mit-founded-startup-sun-catalytix-to-provide-solar-power -storage-for-low-income-houses-in-india [accessed January 18, 2011].

87 *annual energy bill*: Smart Grid Facts, Energy Future Coalition. www
.energyfuturecoalition.org/files/webfmuploads/Transmission%20Smart%20
Grid%20Fact%20Sheet%2002.20.09.pdf [accessed January 22, 2011].

88 *company called Opower*: Amy J. C. Cuddy and Kyle T. Doherty, "Opower:
Increasing Energy Efficiency Through Normative Influence," Harvard
Business School Case Study N2–911–016, November 3, 2010.

90 *to conserve water*: Robert B. Cialdini, "Don't Throw in the Towel: Use
Social Influence Research," *Association for Psychological Science Observer*, April 2005. www.psychologicalscience.org/observer/getArticle
.cfm?id=1762 [accessed January 25, 2011].

91 *level of education*: Michael Watts, "The neighbourhood energy revolution," *Wired*, August 2011. www.wired.co.uk/magazine/archive/2011/08/
features/the-neighbourhood-energy-revolution [accessed September 22,
2011].

91 *changing light bulbs*: Leslie Brooks Suzukamo, "Minnesota gets A+ for
energy report cards," *St. Paul Pioneer Press*, August 13, 2011, A12.

91 *U.S. solar industry*: "Opower to Save One Terawatt Hour of Energy by
2012," press release, June 15, 2011. http://opower.com/company/news
-press/press_releases/25 [accessed September 24, 2011].

CHAPTER 3: THE POWER OF CLUSTERS

93 *In 2008, the United Nation reported*: United Nations Population Division,
"An Overview of Urbanization, Internal Migration, Population Distribution and Development in the World," paper presented at the United Nations Expert Group Meeting on Population Distribution, Urbanization,
Internal Migration and Development, New York, NY, January 21–23, 2008.

94 *a major reurbanization*: The Brookings Institution, *State of Metropolitan
America: On the Front Lines of Demographic Transformation*, Metropolitan
Policy Program at the Brookings Institution, 2010.

96 *in a predictable and systematic way*: Geoffrey B. West, James H. Brown,
and Brian J. Enquist, "A General Model for the Origin of Allometric
Scaling Laws in Biology," *Science* 4 (1997): 122–26.

96 *into her midfifties*: Hillard Kaplan, Kim Hill, Jane Lancaster, and
A. Magdalena Hurtado "A Theory of Human Life History Evolution: Diet,
Intelligence and Longevity," *Evolutionary Anthropology* 9 (4): 156–185.

97 *universal scaling laws*: Luís M. A. Bettencourt, José Lobo, Dirk Helbing,
Christian Kühnert, and Geoffrey B. West, "Growth, Innovation, Scaling
and the Pace of Life in Cities," *PNAS* 17 (2007): 7301–6.

102 *world's most endangered species*: "Primates in Peril: The World's 25 Most Endangered Primates," *Primate Conservation* 24 (2009): 1–57.

102 *50,000 orangutans are left*: S. A. Wich et. al., "Distribution and conservation status of the orang-utan (Pongo spp.) on Borneo and Sumatra: How many remain?" *Oryx* 42 (2008): 329–39.

102 *UN report released in 2007*: C. Nellemann, L. Miles, B. P. Kaltenborn, M. Virtue, and H. Ahlenius (eds.), *The Last Stand of the Orangutan*, United Nations Environment Programme, 2007.

104 *50 percent of all consumer goods*: "Promoting the growth and use of sustainable palm oil," RSPO fact sheet, 2008. www.rspo.org/resource_centre /RSPO_Fact_sheets_Basic.pdf [accessed August 2010].

104 *more than 1,500 percent*: "The other oil spill," *Economist*, June 24, 2010. www.economist.com/node/16423833 [accessed August 15, 2010].

104 *trend is accelerating*: Elizabeth Rosenthal, "Once a Dream Fuel, Palm Oil May Be an Eco-Nightmare," *New York Times*, January 31, 2007. www .nytimes.com/2007/01/31/business/worldbusiness/31biofuel.html [accessed July 25, 2010].

105 *New research estimates*: "Envisat focuses on carbon-rich peat swamp forest fires," European Space Agency website. www.esa.int/esaCP/SEMRA 7YO4HD_index_0.html [accessed July 22, 2010].

105 *total man-made greenhouse gas emissions*: *How the Palm Oil Industry Is Cooking the Climate*, Greenpeace report, November 8, 2007. www.greenpeace .org/usa/en/media-center/reports/how-the-palm-oil-industry-is-c [accessed July 12, 2010].

106 *was an oft-heard refrain*: E. Purwanto and G. A. Limberg, "Global Aspirations to Local Actions: Can Orangutans Save Tropical Rainforest?" paper presented at the Twelfth Biennial Conference of the International Association for the Study of Commons, Cheltenham, England, July 14–18, 2008.

107 *half of the population was without work*: "Willie Smits," profile page, Ashoka.org. http://ashoka.org/fellow/willie-smits [accessed November 30, 2011].

109 *150,000 saplings from his nursery*: Jane Braxton Little, "Regrowing Borneo, Tree by Tree," *Scientific American Earth 3.0*, 18, no. 5 (2008): 64–71.

112 *Smits turned to the tappers of Sulawesi*: "Steaming Ahead," video produced for one of Smits's foundations, Masarang, after it became a finalist in the BBC World Challenge 2007 for charities. www.youtube.com/ watch?v=3_jyN_ASKDE [accessed November 30, 2011].

114 *both fuel and electricity*: Ibid.

115 *bang for the buck*: For a more detailed investigation into this topic, see online conversations between conservationist Erik Meijaard and Willie Smits on the website ConservationBytes. http://conservationbytes.com/2009/07/25/ray-of-conservation-light-for-borneo [accessed November 30, 2011].

116 *what's actually been accomplished*: For a more detailed discussion of criticism from the scientific community, see Little's article "Regrowing Borneo, Tree by Tree."

CHAPTER 4: THE RESILIENT MIND

119 *in a letter in 1946*: Sarah Moskovitz, "Longitudinal Follow-up of Child Survivors of the Holocaust," *Journal of the American Academy of Child Psychiatry* 24, no. 4 (1985): 402.

120 *by Anna Freud and Sophie Dann*: Anna Freud and Sophie Dann, "An Experiment in Group Upbringing," *Psychoanalytic Study of the Child* 6 (1951): 127–68.

121 *"cases involving children"*: Moskovitz, "Longitudinal Follow-up of Child Survivors of the Holocaust," 404.

121 *with schizophrenic patients*: Norman Garmezy and Eliot H. Rodnick, "Premorbid adjustment and performance in schizophrenia: Implications for interpreting heterogeneity in schizophrenia," *Journal of Nervous and Mental Disease* 129 (1959): 450–66.

122 *"to survival and adaptation"*: Norman Garmezy, "Vulnerability Research and the Issue of Primary Prevention," *American Journal of Orthopsychiatry* 41 (1971): 101–16.

122 *"vulnerable but invincible"*: Elwyn James Anthony, "Risk, vulnerability, and resilience: An overview," in *The Invulnerable Child* (New York: Guilford Press, 1987), 3–48. C. Kauffman, H. Grunebaum, B. Cohler, et al., "Superkids: Competent Children of Psychotic Mothers," *American Journal of Psychiatry* 136 (1979): 1398–1402. E. E. Werner and Ruth S. Smith, *Vulnerable but Invincible: A Longitudinal Study of Resilient Children and Youth* (New York: McGraw-Hill, 1982).

122 *"human adaptational systems"*: Ann S. Masten, "Ordinary Magic: Resilience Processes in Development," *American Psychologist* 56 (2001): 227–38.

122 *titled her paper "Ordinary Magic"*: Ibid.

123 *most of all himself*: In addition to his scholarly work, George Bonanno offers a deeper examination of these discoveries in his book *The Other Side of Sadness* (New York: Basic Books, 2009), 1–231.

123 *height of the First World War*: Sigmund Freud, *Mourning and Melancholia,* XVII, 2nd ed. (originally published in 1917; reprinted by Hogarth Press, London, 1955).

124 *a seminal paper*: Eric Lindemann, "Symptomatology and Management of Acute Grief," *American Journal of Psychiatry* 101 (1944): 1141–48.

124 *five stages of mourning*: Elisabeth Kübler-Ross, *On Death and Dying* (New York: Routledge, 1973).

125 *project called Changing Lives of Older Couples*: CLOC Study: Changing Lives of Older Couples: A Multi-Wave Prospective Study of Bereavement. http://www.cloc.isr.umich.edu [accessed December 2, 2011].

125 *their relative distribution across the group*: K. Boerner et al., "Resilient or At-Risk? A Four Year Study of Older Adults Who Initially Showed High or Low Distress Following Conjugal Loss," *Journal of Gerontology: Psychological Sciences* 60B (2005): 67–73.

126 *groups of New Yorkers*: G. A. Bonanno, C. Rennicke, and S. Dekel, "Self-Enhancement Among High-Exposure Survivors of the September 11th Terrorist Attack: Resilience or Social Maladjustment?" *Journal of Personality and Social Psychology* 88, no. 6 (2005): 984–98.

127 *studies on loss and trauma*: G. A. Bonanno et al., "Psychological Resilience After Disaster: New York City in the Aftermath of the September 11th Terrorist Attack," *Psychological Science* 17, 2007 181–186; G. A. Bonanno et al., "What Predicts Resilience After Disaster? The Role of Demographics, Resources, and Life Stress," *Journal of Consulting and Clinical Psychology,* 75, 2007 671–82; G. A. Bonanno et al., "Psychological Resilience and Dysfunction Among Hospitalized Survivors of the SARS Epidemic in Hong Kong: A Latent Class Approach," *Health Psychology* 27 (2008): 659–67.

127 *highly regarded longitudinal study*: J. H. Block and J. Block, "The role of ego-control and ego-resiliency in the organization of behavior," *Development of cognition, affect, and social relations: Minnesota Symposia on Child Psychology,* 13 (1980): 39–101.

127 *this as* hardiness: S. C. Kobasa, "Stressful life events, personality, and health—Inquiry into hardiness," *Journal of Personality and Social Psychology* 37 (1979): 1–11.

128 *religion and resilience*: Kenneth I. Pargament, *The Psychology of Religion and Coping: Theory, Research, Practice* (New York: Guilford, 1997).

128 *"Religion as a Cultural System"*: C. Geertz, *The Interpretation of Cultures: Selected Essays* (New York: Basic Books, 1973), 107–8.

129 *to their Hispanic heritage*: E. Fuentes-Afflick, N. A. Hessol, and E. J. Pérez-Stable, "Testing the epidemiologic paradox of low birth weight in Latinos," *Archives of Pediatrics and Adolescent Medicine* 153 (1999): 147–53; J. B. Gould, A. Madan, C. Qin, and G. Chavez, "Perinatal Outcomes in Two Dissimilar Immigrant Populations in the United States: A Dual Epidemiological Paradox," *Pediatrics* 111 (2003): 676–82.

129 *Ruth S. Smith, published in 2001*: E. E. Werner and R. S. Smith, *Journeys from Childhood to Midlife: Risk, Resilience and Recovery* (Syracuse, N.Y.: Cornell University Press, 2001).

129 *researchers at the University of Maryland, College Park*: S. M. Nettles, W. Mucherah, and D. S. Jones, "Understanding Resilience: The Role of Social Resources," *Journal of Education for Students Placed at Risk,* 5 (2000): 47–60.

130 *Sarah Pressman and Sheldon Cohen*: S. D. Pressman, S. Cohen, G. E. Miller, A. Barkin, B. Rabin, and J. J. Treanor, "Loneliness, Social Network Size, and Immune Response to Influenza Vaccination in College Freshmen," *Health Psychology* 24 (2005): 297–306.

130 *Alexis Stranahan, David Khalil, and Elizabeth Gould*: A. M. Stranahan, D. Khalil, and E. Gould, "Social isolation delays the positive effects of running on adult neurogenesis," *Nature Neuroscience* 9 (2006): 526–33.

130 *researchers from the University of Otago*: "History of the Study," Dunedin Multidisciplinary Health and Development Research Unit. http://dunedinstudy.otago.ac.nz/about-us/how-we-began/history-of-the-study [accessed November 30, 2011].

130 *a gene called* 5-HTT: A. Caspi et al., "Influence of Life Stress on Depression: Moderation by a Polymorphism in the 5-HTT Gene," *Science* 301 (2003): 386–89.

131 *with mice and monkeys*: S. J. Suomi, "Risk, Resilience, and Gene X Environment Interactions in Rhesus Monkeys," *Annals of New York Academy of Sciences* 1094 (2006): 52–62. J. C. Carroll et al., "Effects of mild early life stress on abnormal emotion-related behaviors in 5-HTT knockout mice," *Behavioral Genetics* 37 (2007): 214–22.

132 *smaller than originally found*: Srijan Sen, Margit Burmeister, and Debashis Ghosh, "Meta-analysis of the association between a serotonin transporter promoter polymorphism (5-HTTLPR) and anxiety-related personality traits," *American Journal of Medical Genetics Part B* (2004): 85–89. doi:10.1002/ajmg.b.20158.

132 *optimism and happiness*: Elaine Fox, Anna Ridgewell, and Chris Ash-
     win, "Looking on the bright side: biased attention and the human
     serotonin transporter gene," *Proceedings of the Royal Society B* (March
     2009): 1747–51. doi:10.1098/rspb.2008.1788. Jan-Emmanuel De Neve,
     "Functional Polymorphism (5-HTTLPR) in the Serotonin Transporter
     Gene Is Associated with Subjective Well-Being: Evidence from a
     U.S. Nationally Representative Sample," *Journal of Human Genetics* 56
     (2011): 456–59.

136 *Davidson and his fellow researchers*: A. Lutz, L. L. Greischar, N. B. Rawl-
     ings, M. Ricard, and R. J. Davidson, "Long-term meditators self-induce
     high-amplitude gamma synchrony during mental practice," *Proceedings of
     the National Academy of Sciences* 101 (2004): 16369–73.

136 *he had been trying to investigate for years*: R. Davidson and A. Lutz, "Bud-
     dha's Brain: Neuroplasticity and Meditation," *IEEE Signal Process Maga-
     zine* 25 (2008): 172–176.

137 *larger-than-average hippocampus*: E. Maguire et al., "Navigation-related
     structural change in the hippocampi of taxi drivers," *Proceedings of the
     National Academy of Sciences* 97 (2000): 4398–4403. www.pnas.org/con-
     tent/97/8/4398.full [accessed November 30, 2011].

137 *higher volumes of cortical matter*: C. Gaser and G. Schlaug, "Brain Struc-
     tures Differ Between Musicians and Non-Musicians," *The Journal of Neu-
     roscience* 23 (2003): 9240–45.

137 *led by Dr. Sara Lazar, reported a suggestive finding*: S. Lazar et al., "Mind-
     fulness Practice Leads to Increases in Regional Brain Gray Matter Den-
     sity," *Psychiatry Research: Neuroimaging* 191 (2011): 36–43.

138 *"with meditation practice"*: Quote taken from "Meditation Experience Is
     Associated with Increased Cortical Thickness," *NeuroReport* 16, no. 17
     (2005): 1893–97.

138 *chips away at telomeres*: E. Epel et al., "Accelerated Telomere Shortening
     in Responses to Life Stress," *Proceedings of the National Academy of Sci-
     ences* 101 (2004): 17312–15.

139 *a possible reverse correlation*: T. Jacobs et al., "Intense Meditation Training,
     Immune Cell Telomerase Activity, and Psychological Mediators," *Psycho-
     neuroendocrinology* 36 (2011): 664–81.

140 *tool for pain management*: R. Kalisch et al., "Anxiety Reduction through
     Detachment: Subjective, Physiological, and Neural Effects," *Journal of
     Cognitive Neuroscience* 17 (2005): 874–83.

142 *circumstances like warfare*: E. A. Stanley et al., "Mindfulness-based Mind
     Fitness Training: A Case Study of a High-Stress Predeployment Military

Cohort," *Cognitive and Behavioral Practice* 18 (2011): 566–76; E. A. Stanley and A. P. Jha, "Mind fitness: Improving operational effectiveness and building warrior resilience," *Joint Force Quarterly* 55 (2009): 144–51.

CHAPTER 5: COOPERATION WHEN IT COUNTS

144 *unusual discovery*: H. H. Dale, "On Some Physiological Actions of Ergot," *Journal of Physiology* 34 (1906): 163–206.

145 *chemist Vincent du Vigneaud*: V. du Vigneaud, C. Ressler, et al., "The Synthesis of Oxytocin1." *Journal of the American Chemical Society* 76 (1954): 3115–21.

146 *described the sequence of events*: Andrew Ross Sorkin, "Lehman Files for Bankruptcy; Merrill Is Sold," *New York Times,* September 14, 2011. www.nytimes.com/2008/09/15/business/15lehman.html ?pagewanted=1&sq =lehman%20brothers%20collapse&st=cse&scp=1; [accessed December 4, 2011]; see also Deborah Solomon, Dennis K. Berman, Susanne Craig, and Carrick Mollenkamp, "Ultimatum by Paulson Sparked Frantic End," *Wall Street Journal,* September 15, 2008. http://online.wsj.com/article/ SB122143670579134187.html [accessed December 4, 2011].

146 *of his career "by far"*: Joshua Zumbrun, "Greenspan Says Crisis 'By Far' Worst, Recovery Uneven," Bloomberg, February 23, 2010. www .bloomberg.com/apps/news?pid=newsarchive&sid=a1aLQ51QXlDA& pos=3 [accessed December 4, 2011].

146 *90 percent from a peak at the beginning of 2008*: Jenny Anderson and Andrew Ross Sorkin, "Lehman Said to Be Looking for a Buyer as Pressure Builds," *New York Times,* September 10, 2008. www.nytimes .com/2008/09/11/business/11lehman.html?_r=1&hp&oref=slogin [accessed December 4, 2011].

147 *Paulson told them*: Suzanne McGee, *Chasing Goldman Sachs: How the Masters of the Universe Melted Wall Street* (New York: Random House, 2011), 354.

149 *"Everybody is exposed"*: "Ultimatum by Paulson Sparked Frantic End," *Wall Street Journal.*

151 *"where does it end?"*: Ibid.

151 *neurobiology of trust*: P. J. Zak, R. Kurzban, et al., "The Neurobiology of Trust," *Annals of the New York Academy of Sciences* 1032 (2004): 224–27; M. Kosfeld, M. Heinrichs, et al., "Oxytocin increases trust in humans," *Nature* 435 (2005): 673–76; Paul J. Zak, "The Neurobiology of Trust," *Scientific American,* June 2008, 88.

152 *oxytocin formulated as a nasal spray*: Kosfeld, Heinrichs, et al., "Oxytocin increases trust in humans."

152 *"through interpersonal interactions"*: Ibid., 673.

152 *"bankers and speculators"*: A. C. Grayling, "Beware the Destructive Nature of Greed," *New Scientist*, November 5, 2008.

153 *intensely anxious social situations*: Zak, "The Neurobiology of Trust."

154 *"if it wasn't happening to us"*: David Cho and Neil Irwin, "No Bailout: Fed Made New Policy Clear in One Dramatic Weekend," *Washington Post*, September 26, 2008. www.washingtonpost.com/wp-dyn/content /article/2008/09/15/AR2008091503312.html?sid=ST2008091503351 &s_pos= [accessed December 4, 2011].

154 *Wall Street had its worst day in seven years*: Patrick Rizzo and Joe Bel Bruno, "Financial Crisis as Dow Drops 504 Points," The Associated Press, September 15, 2008. www.seattlepi.com/business/article/Financial -crisis-as-Dow-drops-504-points-1285321.php#page-1 [accessed December 4, 2011].

157 *known as* kin selection: W. D. Hamilton, "The genetical evolution of social behaviour, I," *Journal of Theoretical Biology* 7 (1964): 1–16.

158 *reciprocal altruism*: R. L. Trivers, "The evolution of reciprocal altruism," *Quarterly Review of Biology* 46 (1971): 35–57.

159 *computer tournament*: R. Axelrod, *The Evolution of Cooperation* (New York: Basic Books, 1985): 3–27.

161 *version of the game*: Drew Fudenberg, David Rand and Anna Dreber, "Slow to Anger and Fast to Forgive: Cooperation in an Uncertain World," *American Economic Review*. In press.

162 *inequity among capuchin monkeys*: S. Brosnan and F. B. M. de Waal, "Monkeys Reject Unequal Pay," *Nature* 425 (2003): 297–99.

163 *"getting a better deal"*: Frans de Waal, "Frans de Waal Answers Your Primate Questions," Freakonomics blog, May 7, 2008. www.freakonomics .com/2008/05/07/frans-de-waal-answers-your-primate-questions [accessed December 3, 2011].

164 *"still buy Lehman"*: Solomon, Berman, Craig, and Mollenkamp, "Ultimatum by Paulson Sparked Frantic End."

164 *"foreign buyer was even less likely"*: Ibid.

164 *University of Amsterdam, believe so*: C. K. W. de Dreu, L. L. Greer, et al., "Oxytocin promotes human ethnocentrism," *PNAS* 108 (2011): 1262–1266. www.pnas.org/content/early/2011/01/06/1015316108.full .pdf [accessed December 4, 2011].

164 *similar hormones at work*: S. Muzafer, O. J. Harvey, B. J. White, W. R. Hood, and C. W. Sherif, *The Robbers Cave Experiment: Intergroup Conflict and Cooperation* (Norman: University of Oklahoma Press, 1961).

164 *mathematically modeled*: S. Rytina and D. L. Morgan, "The Arithmetic of Social Relations: The Interplay of Category and Network," *American Journal of Sociology* 88 (1982): 88–113.

166 *"to whom the codes apply"*: Edward O. Wilson, *On Human Nature* (Cambridge, Mass.: Harvard University Press, 1978): 163.

166 *"forth with ease"*: Ibid.

166 *participate in an unusual experiment*: S. C. Wright et al., "The extended contact effect: Knowledge of cross-group friendships and prejudice," *Journal of Personality and Social Psychology* 73 (1997): 73–90; see also the Interpersonal Relationships Lab at SUNY Stony Brook: www.psychology .stonybrook.edu/aronlab- [accessed December 2, 2011].

167 *that is as close as any in a person's life*: Benedict Carey, "Tolerance over Race Can Spread, Studies Find," *New York Times*, November 6, 2008. www.nytimes.com/2008/11/07/us/07race.html?scp=2&sq=%22art%20 aron%22&st=cse [accessed December 4, 2011].

173 *316,000 Haitians were killed*: US Geological Survey. http://earthquake .usgs.gov/earthquakes/eqinthenews/2010/us2010rja6/#summary [accessed December 4, 2011]

173 *300,000 were injured*: "Haiti quake death toll rises to 230,000," *BBC News*. February 11, 2010. http://news.bbc.co.uk/2/hi/8507531.stm [accessed December 4, 2011]

173 *million more were made homeless*: "Haiti will not die, President Rene Preval insists," *BBC News*, February 12, 2010. http://news.bbc.co.uk/2/hi/americas /8511997.stm [accessed December 4, 2011]

173 *three million in all were affected*: "Red Cross: 3M People Affected by Quake," *CBS News*, March 9, 2010 http://www.cbsnews.com/stories /2010/01/13/world/main6090601.shtml?tag=cbsnewsSectionContent.4 [accessed December 4, 2011].

185 *"open source world"*: "Some positive feedback," Mission 4636 blog, February 10, 2010. www.mission4636.org/some-positive-feedback [accessed December 4, 2011].

185 *reports that the average response time*: Ryan Ferrier, "Crowdsourcing the Haiti Relief: One Year Later," CrowdFlower blog, January 11, 2011. http://blog.crowdflower.com/2011/01/crowdsourcing-the-haiti-relief -one-year-later/ [accessed December 4, 2011].

186 *Stanford sociologist Mark Granovetter*: M. S. Granovetter, "The Strength of Weak Ties," *American Journal of Sociology* 78 (1973): 1360–80.

187 *weak ties may be incomplete*: S. Aral and M. V. Alstyne, "Networks, Information and Brokerage: The Diversity-Bandwidth Trade-off," April 15, 2010. Forthcoming in the *American Journal of Sociology*.

189 *large-scale disaster response*: Disaster Relief: The Future of Information Sharing in Humanitarian Emergencies. United Nations Foundation Report. www.unfoundation.org/news-and-media/publications-and-speeches /disaster-relief-2-report.html [accessed December 4, 2011]

189 single *list of schools*: Ibid.

CHAPTER 6: COGNITIVE DIVERSITY

191 *Willoughby Verner*: Letter quoted in John Adams's *Risk* (London: UCL Press, 1995), 113.

193 *not decrease in number*: D. Albury and J. Schwartz, *Partial Progress* (London: Pluto Press, 1982), 9–24.

193 *imposed throughout the late 1960s*: S. Peltzman, "The effects of automobile safety regulation," *Journal of Political Economy* 83 (1975): 677–726.

193 *impact of seat belts on highway fatalities*: J. Adams, "The efficacy of seatbelt legislation: A comparative study of road accident fatality statistics from 18 countries," Department of Geography, Occasional Paper, University College, London (1981).

193 *many essays on risk*: J. Adams, "Seat Belts—Blood on My Hands?" Blog post on John Adams's website, March 5, 2008. www.john-adams.co.uk/2008/03 /05/seat-belts—blood-on-my-hands [accessed December 2, 2011].

194 *a tendency to take more physical risks*: B. A. Morrongiello, B. Walpole, and J. Lasenby, "Understanding children's injury-risk behavior: Wearing safety gear can lead to increased risk taking," *Accident Analysis and Prevention* 39 (2007): 618–23.

194 *riskier sexual behavior*: M. Cassell et al., "Risk compensation: The Achilles' Heel of Innovations in HIV Prevention," *BMJ* 332 (2006): 332.

194 *rescuer can access them easily*: "Climber 9–1–1," *Northwest Mountaineering Journal*. www.mountaineers.org/nwmj/10/101_Rescue2.html [accessed December 2, 2011].

194 *"fatality rate constant"*: V. Napier, "Risk Homeostasis: A Case Study of the Adoption of a Safety Innovation on the Level of Perceived Risk," Vic Napier's website. www.vicnapier.com/MyArticles/Parachutes_Skydiving /skydivers_risktaking_behavior.htm [accessed December 2, 2011].

194 *analogous to a thermostat*: Adams, *Risk;* see also G. J. S. Wilde, "Critical Issues in Risk Homeostasis Theory," *Risk Analysis* 2 (1982): 249–58.

194 *"no one wants absolute zero"*: Adams, *Risk,* 15.

195 *three-year study of taxicabs in Munich*: G. J. S. Wilde, *Target Risk: Dealing with the Danger of Death, Disease and Damage in Everyday Decisions* (Toronto: PDE Publications, 1994), chapter 7.1.

196 *"not protected by the caps"*: W. K. Viscusi, "Consumer Behavior and the Safety Effects of Product Safety Regulation," *Journal of Law and Economics* 28 (1985): 527–53.

197 *"culture could exist at BP"*: Quote taken from a speech given at the National Petrochemical and Refiners Association conference in San Antonio, Texas, on March 19, 2007.

197 *resistant to Hayward's beam*: Jad Mouawad, "For BP, a History of Spills and Safety Lapses," *New York Times,* May 8, 2010. www.nytimes .com/2010/05/09/business/09bp.html?pagewanted=all [accessed November 29, 2011].

198 *expressed grave concerns*: Scott Bronstein and Wayne Drash, "Rig survivors: BP ordered shortcut on day of the blast," CNN website, June 8, 2010. http://articles.cnn.com/2010–06–08/us/oil.rig.warning.signs_1 _rig-transocean-bp?_s=PM:US [accessed December 2, 2011].

198 *"maintaining their complex facilities"*: From conservativehome.blogs .com, June 3, 2010. http://conservativehome.blogs.com/platform/2010/06 /oberon-houston-.html [accessed December 2, 2011].

198 *in a 2009 survey*: J. M. Farrell and A. Hoon, "What's Your Company's Risk Culture?" National Association of Corporate Directors Directorship, April 15, 2009. www.kpmg.com/MT/en/IssuesAndInsights/Articles Publications/Documents/Risk-culture.pdf [accessed December 2, 2011].

204 *Page points to the diversity prediction theorem*: Scott Page, *The Difference: How the Power of Diversity Creates Better Groups, Firms, Schools and Societies* (Princeton, N.J.: Princeton University Press, 2007), 197–39.

205 *experimentally validated by Kevin Dunbar*: K. Dunbar, "How scientists really reason: Scientific reasoning in real-world laboratories," in R. J. Sternberg and J. Davidson (eds.), *Mechanisms of Insight.* (Cambridge, Mass.: MIT Press, 1995), 365–95.

206 *"Lab A"*: Dunbar keeps the identities of the labs and personnel confidential, to protect the professional reputations involved. We've kept the description generic, in keeping with his wishes.

207 *research by Sinan Aral at New York University*: S. Aral and M. V. Alstyne, "Networks, Information and Brokerage: The Diversity-Bandwidth

Trade-off," April 15, 2010. Forthcoming in the *American Journal of Sociology*.

208 *psychologist Peter Wason in 1960*: P. Wason, "On the failure to eliminate hypotheses in a conceptual task," *Quarterly Journal of Experimental Psychology* 12 (1960): 129–40.

209 *found that when watching* The Colbert Report: H. L. LaMarre et al., "The Irony of Satire: Political Ideology and the Motivation to See What You Want to See in *The Colbert Report*," *International Journal of Press/Politics* 14 (2009): 212–31.

CHAPTER 7:
COMMUNITIES THAT BOUNCE BACK

212 *cholera, typhoid, and hepatitis*: Andrew Meharg. *Venomous Earth: How Arsenic Caused the World's Worst Mass Poisoning* (London: Palgrave Macmillan, 2004), 6.

212 *in 2010 combined*: Julie Reed Bell and Seth Borenstein, "2010's world gone wild: Quakes, floods, blizzards," Associated Press, December 19, 2010. www.msnbc.msn.com/id/40739667/ns/us_news-2010_year_in_review/t/s-world-gone-wild-quakes-floods-blizzards/#.TtsPJXOfvh9 [accessed December 3, 2011].

212 *sunk with UNICEF's assistance*: Maggie Black, *The Children and the Nations: The Story of UNICEF* (New York: UNICEF, 1986), 301.

212 *water by the year 2000*: Meharg, *Venomous Earth*, 7.

212 *less than 10 percent in the late 1990s*: *Fighting Human Poverty: Bangladesh Human Development Report 2000* (UNDP: 2000) http://hdr.undp.org/en/reports/national/asiathepacific/bangladesh/name,2748,en.html [accessed December 1, 2011].

213 *included in a new bride's dowry*: Meharg, *Venomous Earth*, 13.

213 *see something disturbing*: Ibid., 12.

213 *skin, lung, and bladder cancer*: "Arsenic Mitigation in Bangladesh," UNICEF Fact Sheet (October 12, 2008). www.unicef.org/bangladesh/media_2121.htm [accessed Nov 4, 2011]; see also United Nations Foundation, *Arsenic Poisoning in Bangladesh and West Bengal*, U.N. Foundation Report, October 1999, 1–20.

213 *an arsenic-related cancer*: "Water Related Diseases: Arsenicosis," World Health Organization. www.who.int/water_sanitation_health/diseases/arsenicosis/en [accessed December 2, 2011].

214 *"our program will go on"*: Barry Bearak, "New Bangladesh Disaster: Wells That Pump Poison," *New York Times*, November 11, 1998 www.nytimes .com/1998/11/10/world/death-by-arsenic-a-special-report-new-bangladesh -disaster-wells-that-pump-poison.html?pagewanted=all&src=pm [accessed on December 2, 2011].

214 *show up in one's food*: Based on an email and phone conversation with Andrew Meharg regarding his research, February 16, 2011.

214 *one out of every two wells*: "Arsenic Mitigation in Bangladesh."

214 *"largest mass poisoning"*: "Researchers Warn of Impending Disaster from Mass Arsenic Poisoning," press release, World Health Organization, September 8, 2000. www.who.int/inf-pr-2000/en/pr2000–55.html [accessed December 2, 2011].

214 *sense of disgust*: S. Hanchett, "Social Aspects of the Arsenic Contamination of Drinking Water," in *Selected Papers on the Social Aspects of Arsenic and Arsenic Mitigation in Bangladesh* (Dhaka: Arsenic Policy Support Unit, Government of Bangladesh, 2006), 2.

215 *country's 86,000 villages*: Bangladesh Arsenic Mitigation Water Supply Program (BAMWSP). www.bamwsp.org [accessed December 10, 2010].

215 *all over their skin*: F. Sultana, "Gender Concerns in Arsenic Mitigation in Bangladesh," in *Selected Papers on the Social Aspects of Arsenic and Arsenic Mitigation in Bangladesh* (Dhaka. Arsenic Policy Support Unit: Government of Bangladesh, 2006), 53–84.

215 *most successful public health interventions in history*: E. Field, R. Glennerster, and R. Hussam, "Throwing the Baby Out with the Drinking Water: Unintended Consequences of Arsenic Mitigation Efforts in Bangladesh," February 14, 2011. http://web.mit.edu/j-pal/www/book/Arsenic_Infant Mortality_feb10.pdf [accessed December 4, 2011].

216 *front of the mosque*: Sultana, "Gender Concerns in Arsenic Mitigation in Bangladesh," 68.

216 *their water resource use*: Ibid., 53–84.

216 *"in his own home?"*: Ibid., 64.

217 *"understood" the status of their well*: Field, Glennerster, and Hussam, "Throwing the Baby Out with the Drinking Water," 2.

217 *difficult to grasp*: Sultana, "Gender Concerns in Arsenic Mitigation in Bangladesh," 69.

217 *curses from God*: Hanchett, "Social Aspects of the Arsenic Contamination of Drinking Water," 13.

217 *Why bring in a sick girl?*: *Selected Papers on the Social Aspects of Arsenic and Arsenic Mitigation in Bangladesh* (Dhaka, Bangladesh: Arsenic Policy Support Unit, Government of Bangladesh, 2006), 1–92.

217 *water options by the government*: *Towards an Arsenic Safe Environment*, a joint Publication of FAO, UNICEF, WHO and WSP, March 2010. www.unicef.org/bangladesh/knowledgecentre_6131.htm [accessed January 29, 201]

218 *fecal pathogens*: K. Lokuge, et al., "The effect of arsenic mitigation interventions on disease burden in Bangladesh," *Environmental Health Perspectives* 112 (2004): 1172.

218 *risk of diarrheal disease by 20 percent*: Ibid.

218 *the highest rate of switching recorded*: A. Schoenfeld, "Area, Village and Household Response to Arsenic Testing and Labeling of Tubewells in Araihazar, Bangladesh," Masters Thesis, Columbia University, September 6, 2005. www.ldeo.columbia.edu/~avangeen/arsenic/documents/Schoenfeld_MS_05.pdf [accessed December 4, 2011].

218 *excessive quantities of arsenic*: *Towards an Arsenic Safe Environment*.

218 *"water for all by 2011"*: Ibid., 4.

220 *Davion*: Name changed to protect the client's anonymity.

223 *during the same period*: Bob McCarty, "2008: Chicago Murders Total Tops U.S. Soldier Deaths in Iraq," Now Public website, January 5, 2009. www.nowpublic.com/world/2008-chicago-murders-total-tops-u-s-soldier-deaths-iraq [accessed December 4, 2011].

225 *the first official document on the disease*: Centers for Disease Control, "Pneumocystis pneumonia—Los Angeles," *Morbidity and Mortality Weekly Report* (1981): 250–52. www.cdc.gov/mmwr/preview/mmwrhtml/lmrk077.htm [accessed December 1, 2011].

225 *had passed away*: Sharon Block, "25 Years of AIDS: June 5, 1981—June 5, 2006," University of California San Francisco Hospital website, June 5, 2006. www.ucsf.edu/news/2006/06/6955/25-years-aids-june-5-1981-june-5-2006 [accessed December 4, 2011].

225 *reported at San Francisco General*: Ibid.

226 *transmission of HIV/AIDS*: I. Rosenstock, V. Strecher, and M. Becker, "The Health Belief Model and HIV risk behavior change," in R. J. DiClemente and J. L. Peterson (eds.), *Preventing AIDS: Theories and Methods of Behavioral Interventions* (New York: Plenum Press, 1994), 5–24.

226 *behaviors of those around them*: M. Fishbein and S. E. Middlestadt, "Using the theory of reasoned action as a framework for understanding and changing AIDS-related behaviors," in V. M. Mays, G. W. Albee, and

S. F. Schneider (eds.), *Primary Prevention of AIDS: Psychological Approaches* (London: Sage Publications, 1989), 93–110.

227 *would use a condom*: M. J. VanLandingham et al., "Two views of risky sexual practices among Northern Thai males : The Health Belief Model and the Theory of Reasoned Action," *Journal of Health and Social Behavior* 36 (1995), 195–212.

227 *from their partners*: S. E. Middlestadt and M. Fishbein, "Factors influencing experienced and inexperienced college women's intentions to tell their partners to use condoms," paper presented at the International Conference AIDS at University of Illinois at Urbana-Champaign, June 20–23, 1990.

233 *and financial resources to go anywhere*: Pam Belluck, "End of a Ghetto," *New York Times*, September 6, 1998. www.nytimes.com/1998/09/06 /us/end-of-a-ghetto-a-special-report-razing-the-slums-to-rescue-the -residents.html?pagewanted=all&src=pm [accessed December 2, 2011].

237 *led by biophysicist Allison Hill*: A. Hill, D. G. Rand, M. A. Nowak, and N. A. Christakis, "Emotions as infectious diseases in a large social network: he SISa model," Proceedings of the Royal Society B: Biological Sciences, published online before print, July 7, 2010. http://rspb.royalsocietypublishing .org/content/early/2010/07/03/rspb.2010.1217 [accessed December 4, 2011].

238 *across all of the communities*: *CeaseFire Evaluation Report*, Institute for Policy Research at Northwestern University (2008) www.ipr.northwestern .edu/publications/ceasefire.html [accessed December 4, 2011].

238 *West and South Side Chicago*: Data and Research, CeaseFire website. http://ceasefirechicago.org/data-research [accessed December 4, 2011].

CHAPTER 8: THE TRANSLATIONAL LEADER

240 *"Understanding Palau"*: The context for understanding the history of Palau's fishing culture would not have been possible without Robert Earle Johannes's rich ethnography *Words of the Lagoon: Fishing and Marine Lore in the Palau District of Micronesia* (Berkeley: University of California Press, 1981).

242 *"operate by himself"*: Johannes, *Words of the Lagoon*, 17.

243 *tinned mackerel salted and shipped in from Japan*: Ibid.

256 *rural communities in Appalachian Ohio*: V. Kerbs and J. Holley, "Building Smart Communities Through Network Weaving," 2002–2006; and "Building Sustainable Communities Through Network Building," 2002. http://www.orgnet.com/cases.html [accessed December 4, 2011].

CHAPTER 9:
BRINGING RESILIENCE HOME

262 *and excessive rain*: "Fact sheet: Kilimo Salama," Syngenta Foundation. www.syngentafoundation.org/__temp/Kilimo_Salama_Fact_sheet _FINAL.pdf [accessed November 10, 2011].

262 *5 percent of the cost*: Ibid.

263 *yield increased 150 percent*: Rose Goslinga, speech at the 2011 PopTech conference. http://poptech.org/popcasts/rose_goslinga_farmer_microinsurance [accessed November 10, 2011].

264 *of whom are currently served*: Jim Roth, Michael J. McCord, and Dominic Liber, *The Landscape of Microinsurance in the World's Poorest 100 Countries* (The MicroInsurance Centre, 2007).

264 *an adhocracy*: See Alvin Toffler, *Future Shock* (New York: Bantam Books, 1970) and Henry Mintzberg, *The Structuring of Organizations: A Synthesis of the Research* (Englewood Cliffs, N.J.: Prentice-Hall, 1979).

266 *sanitation projects over time*: www.waterforpeople.org/programs/field -level-operations-watch.html [accessed July 10, 2011].

267 *EpidemicIQ*: http://epidemiciq.com [accessed November 5, 2011].

268 ahead of the *World Health Organization*: Personal correspondence with Rob Munro, chief technology officer, EpidemicIQ.

269 *as pothole free as possible*: http://open311.org [accessed July 14, 2010].

269 *"Pay attention."*: John Burnett, "Mexican Cartels Open New Front in War: Online," National Public Radio, November 18, 2011. www.npr .org/2011/11/18/142518965/mexican-cartels-open-new-front-in-war -online [accessed December 4, 2011]

270 *compatriots shot him*: Ashish Khetan, "60 Dark Hours at Hotel Taj," in *26/11 Mumbai Attacked*, H. Baweja (ed.), (New Delhi: Roli Books. 2009), pp. 46–83.

270 *Natural Capital Project*: www.naturalcapitalproject.org/marine/Marine InVEST_Apr2010.pdf [accessed August 1, 2010].

270 *live updates about the attack*: H. Raghav Rao, "Beyond Information Assurance: Information Control and Terrorism," *http://icmis.iiita.ac.in/ppt/21/ hrrao21_1.ppt* [accessed December 4, 2011].

273 *seventeen-story headquarters*: James Pat Smith, *Leadership and Mission in Resilient Organizations: Hancock Bank as a Case Study*, Community and Regional Resilience Institute. www.resilientus.org/library/GP _Resilience_Essay_Hancock_Bank_Final_8409_1249429792.pdf [accessed September 23, 2011].

273 *IDs and checkbooks*: Ken Belson, "After Hurricane Katrina, a Bank Turns to Money Laundering," *New York Times*, September 29, 2005. www .nytimes.com/2005/09/29/business/29hancock.html [accessed September 23, 2011].

274 *had been paid back*: Smith, *Leadership and Mission in Resilient Organizations*, 3.

INDEX

ABOUT THE AUTHORS

ANDREW ZOLLI is a renowned expert in three connected fields: foresight, social innovation, and resilience. He serves as the curator and executive director of PopTech (*www.poptech.org*), a global innovation network that brings together eminent and emerging leaders in science, technology, innovation, design, health, the humanities, and the corporate and social sectors to collaborate on breakthrough new approaches to the world's toughest challenges.

Under his leadership, PopTech has developed pioneering programs that identify and train some of the world's most impactful social innovators and scientists, and spurred significant advances in mobile healthcare, education, climate adaptation, and a number of related fields. PopTech's annual thought leadership conference, held annually in Camden, Maine, has become one of the most influential innovation-focused events in the United States.

Andrew has also served as a fellow of the National Geographic Society, and his work, writings, and ideas have appeared in a wide array of media outlets, including *The New York Times, Wired, PBS, National Public Radio, BusinessWeek, Fast Company, Newsweek,* and many others.

He feels very lucky to live in Brooklyn, New York.

ANN MARIE HEALY is a playwright, screenwriter, and journalist. In addition to receiving numerous productions across the country, her plays are also anthologized in collections from Smith & Kraus and Samuel French and featured in journals such as *The Kenyon Review* and *Play: A Journal of Plays.* She currently lives in the Hudson River Valley.